Lecture Notes in Earth Sciences

55

Editors:

S. Bhattacharji, Brooklyn
G. M. Friedman, Brooklyn and Troy
H. J. Neugebauer, Bonn
A. Seilacher, Tuebingen and Yale

W0050133

Springer-Verlag Berlin Heidelberg GmbH

John A. Scales

Theory of
Seismic Imaging

 Springer

Author

John A. Scales
Center for Wave Phenomena
Department of Geophysics, Colorado School of Mines
Golden, Colorado 80401, USA

Cover illustration: The Marmousi Model

This is a model of subsurface geology created by the Institut Francais du Pétrole in
1988 in order to test state-of-the-art migration velocity analysis methods. It is based
on a profile in the Cuanza basin (Angola). For more details, see Analyse du problème
de la détermination du modèle de vitesse pour l'imagerie sismique by Versteeg and
Grau (Eds.), (Thèse de Doctorate de l'Université Paris VII, EAEG Zeist 1991).

ISBN 978-3-540-59051-4   ISBN 978-3-540-49180-4 (eBook)
DOI 10.1007/978-3-540-49180-4
CIP data applied for

This work is subject to copyright. All rights are reserved, whether the whole or part
of the material is concerned, specifically the rights of translation, reprinting, re-use
of illustrations, recitation, broadcasting, reproduction on microfilms or in any other
way, and storage in data banks. Duplication of this publication or parts thereof is
permitted only under the provisions of the German Copyright Law of September 9,
1965, in its current version, and permission for use must always be obtained from
Springer-Verlag Berlin Heidelberg GmbH. Violations are liable for prosecution under
the German Copyright Law.

© Springer-Verlag Berlin Heidelberg 1995
Originally published by Springer-Verlag Berlin Heidelberg New York in 1995

Typesetting: Camera ready by author
SPIN: 10492893   32/3142-543210 - Printed on acid-free paper

Preface

Seismic imaging is the process through which seismograms recorded on the Earth's surface are mapped into representations of its interior properties. Imaging methods are nowadays applied to a broad range of seismic observations: from near-surface environmental studies, to oil and gas exploration, even to long-period earthquake seismology. The characteristic length scales of the features imaged by these techniques range over many orders of magnitude. Yet there is a common body of physical theory and mathematical techniques which underlies all these methods.

The focus of this book is the imaging of reflection seismic data from controlled sources. At the frequencies typical of such experiments, the Earth is, to a first approximation, a vertically stratified medium. These stratifications have resulted from the slow, constant deposition of sediments, sands, ash, and so on. Due to compaction, erosion, change of sea level, and many other factors, the geologic, and hence elastic, character of these layers varies with depth and age. One has only to look at an exposed sedimentary cross section to be impressed by the fact that these changes can occur over such short distances that the properties themselves are effectively discontinuous relative to the seismic wavelength. These layers can vary in thickness from less than a meter to many hundreds of meters. As a result, when the Earth's surface is excited with some source of seismic energy and the response recorded on seismometers, we will see a complicated zoo of elastic wave types: reflections from the discontinuities in material properties, multiple reflections within the layers, guided waves, interface waves which propagate along the boundary between two different layers, surface waves which are exponentially attenuated with depth, waves which are refracted by continuous changes in material properties, and others. The character of these seismic waves allows seismologists to make inferences about the nature of the subsurface geology.

Because of tectonic and other dynamic forces at work in the Earth, this first-order view of the subsurface geology as a layer cake must often be modified to take into account bent and fractured strata. Extreme deformations can occur in processes such as mountain building. Under the influence of great heat and stress, some rocks exhibit a taffy-like consistency and can be bent into exotic shapes without breaking, while others become severely fractured. In marine environments, less dense salt can be overlain by more dense sediments; as the salt rises under its own buoyancy, it pushes the overburden out of the way, severely deforming originally flat layers. Further, even on the relatively localized scale of exploration seismology, there may be significant lateral variations in material properties. For example, if we look at the sediments carried downstream by a river, it is clear that lighter particles will be carried further, while bigger ones will be deposited first; flows near the center of the channel will be faster than the flow on the verge. This gives rise to significant variation is the density and porosity of a given sedimentary formation as a function of just how the sediments were deposited.

Taking all these effects into account, seismic waves propagating in the Earth will be refracted, reflected and diffracted. In order to be able to image the Earth, to see through the complicated distorting lens that its heterogeneous subsurface presents to us, in other words, to be able to solve the inverse scattering problem, we need to be able to undo all of these wave propagation effects. In a nutshell, that is the goal of imaging: to transform a suite of seismograms recorded at the surface of the Earth into a *depth section*, i.e., a spatial image of some property of the Earth (usually wave speed or impedance). There are two main types of spatial variations of the Earth's properties. There are the smooth changes (smooth meaning possessing spatial wavelengths which are long compared to seismic wavelengths) associated with processes such as compaction. These gradual variations cause ray paths to be gently turned or refracted. On the other hand, there are the sharp changes (short spatial wavelength), mostly in the vertical direction, which we associate with changes in lithology and, to a lesser extent, fracturing. These short wavelength features give rise to the reflections and diffractions we see on seismic sections. If the Earth were only smoothly varying, with no discontinuities, then we would not see any events at all in exploration seismology because the distances between the sources and receivers are not often large enough for rays to turn upward and be recorded. This means that to first order, reflection seismology is sensitive primarily to the short spatial wavelength features in the velocity model. We usually assume that we know the smoothly varying part of the velocity model (somehow) and use an imaging algorithm to find the discontinuities.

The earliest forms of imaging involved moving, literally migrating, events around seismic time sections by manual or mechanical means. Later, these manual migration methods were replaced by computer-oriented methods which took into account, to varying degrees, the physics of wave propagation and scattering. It is now apparent that all accurate imaging methods can be viewed essentially as linearized inversions of the wave equation, whether in terms of Fourier integral operators or direct gradient-based optimization of a waveform misfit function. The implicit caveat hanging on the word "essentially" in the last sentence is this: people in the exploration community who practice migration are usually not able to obtain or preserve the true amplitudes of the data. As a result, attempts to interpret subtle changes in reflector strength, as opposed to reflector position, usually run afoul of one or more approximations made in the sequence of processing steps that makes up a migration (trace equalization, gaining, deconvolution, etc.) On the other hand, if we had true amplitude data, that is, if the samples recorded on the seismogram really were proportional to the velocity of the piece of Earth to which the geophone were attached, then we could make quantitative statements about how spatial variations in reflector strength are related to changes in geological properties. The distinction here is the distinction between imaging reflectors, on the one hand, and doing a true inverse problem for the subsurface properties on the other.

Until quite recently the exploration community was exclusively concerned with the former, and today the word "migration" almost always refers to the imaging problem. The more sophisticated view of imaging as an inverse problem is gradually making its way into the production software of oil and gas exploration companies, since careful treatment of amplitudes is often crucial in making decisions on subtle lithologic plays (amplitude versus offset or AVO) and in resolving the chaotic wave propagation effects of complex structures.

When studying migration methods, the student is faced with a bewildering assortment of algorithms, based upon diverse physical approximations. What sort of velocity model can be used: constant wave speed v? $v(x)$, $v(x, z)$, $v(x, y, z)$? Gentle dips? Steep dips? Shall we attempt to use turning or refracted rays? Take into account mode converted arrivals? 2D (two dimensions)? 3D? Prestack? Poststack? If poststack, how does one effect one-way wave propagation, given that stacking attenuates multiple reflections? What domain shall we use? Time-space? Time-wave number? Frequency-space? Frequency-wave number? Do we want to image the entire dataset or just some part of it? Are we just trying to refine a crude velocity model or are we attempting to resolve an important feature with high resolution? It is possible to imagine imaging algorithms that would work under the most demanding of these assumptions, but they would be highly inefficient when one of the simpler physical models pertains. And since all of these situations arise at one time or another, it is necessary to look at a variety of migration algorithms in daily use.

Given the hundreds of papers that have been published in the past 15 years, to do a reasonably comprehensive job of presenting all the different imaging algorithms would require a book many times the length of this one. This was not my goal in any case. I have tried to emphasize the fundamental physical and mathematical ideas of imaging rather than the details of particular applications. I hope that rather than appearing as a disparate bag of tricks, seismic imaging will be seen as a coherent body of knowledge, much as optics is.

Acknowledgement. I would like to thank my colleagues at the Center for Wave Phenomena for their support and encouragement in the preparation of this book; especially Jack Cohen and John Stockwell, who contributed the appendix on SU and offered much good advice on the mathematical aspects of wave phenomena, Ken Larner, who corrected many mistakes in the first draft, and Barbara McLenon, who typed parts of Chapter 11 from some old notes of mine. Special thanks go to Martin Smith of New England Research, who wrote the appendix on SUB and co-authored the chapters on elastic waves, but whose biggest contributions to this book remain undocumented. I am grateful to all my colleagues who spoke about their research before the class, including Christof Stork, Paul Fowler, Mike Sullivan, Craig Artley and Paul Docherty. These talks have become an integral part of the course. Paul Docherty also created the synthetic dataset used in the final computer exercise. Finally, my wife Pamela, without whom none of this would have been possible, deserves a medal for putting up with me.

Table of Contents

1. Introduction to Seismic Migration

1.1 The Reflection Seismic Experiment

The essential features of an exploration seismic experiment are:

- Using controlled sources of seismic energy

- Illumination of a subsurface target area with the downward propagating waves

- Reflection, refraction, and diffraction of the seismic waves by subsurface heterogeneities

- Detection of the backscattered seismic energy on seismometers spread out along a linear or areal array on the Earth's surface.

On land, the seismometers are called geophones. Generally they work by measuring the motion of a magnet relative to a coil attached to the housing and implanted in the Earth. This motion produces a voltage which is proportional to the velocity of the Earth. Most geophones in use measure just the vertical component of velocity. However, increasing use is being made of multicomponent receivers in order to be able to take advantage of the complete elastic wavefield. For marine work, hydrophones are used which sense the instantaneous pressure in the water due to the seismic wave.

The receivers are deployed in clusters called groups; the signal from each receiver in a group is summed so as to a) increase the signal to noise ratio, and b) attenuate horizontally propagating waves. On land, horizontally-propagating Rayleigh waves are called ground roll and are regarded by explorationists as a kind of noise inasmuch as they can obscure the sought-after reflection events. The reason that receiver group summation attenuates horizontally propagating signals and not the vertically propagating reflected events is that the ray paths for the vertically propagating events strike the receivers at essentially the same time, whereas if the receivers in a group are placed so that the horizontally propagating wave is out of phase on the different receivers, then summing these signals will result in cancellation. Sometimes the receivers are buried in the ground if the environment is particularly noisy–in fact, burying the receivers is always a good idea, but it's too expensive and time-consuming in all but special cases. The individual receiver groups are separated from one another by distances of anywhere from a few dozen meters to perhaps 100 meters. The entire seismic line will be several kilometers or more long.

Seismic sources come in all shapes and sizes. On land these include, for example, dynamite, weight drops, large caliber guns, and large resistive masses called vibrators, which are excited with a chirp or swept continuous-wave signal. For marine surveys, vibrators, air guns, electric sparkers, and confined propane-oxygen explosions are the most common sources. The amount of dynamite used can range from a few grams for near-surface high-resolution studies, to tens of kilograms or more. The amount used depends on the type of rock involved and the depth of the target. Explosive charges require some sort of hole to be drilled to contain the blast. These holes are often filled with heavy mud in order to help project the energy of the blast downward. Weight drops involve a truck with a crane, from which the weight is dropped several meters. This procedure can be repeated quickly. Small dynamite blasts (caps) and large caliber guns (pointed down!) are popular sources for very high-resolution near-surface studies. By careful stacking and signal processing, the data recorded by a vibrating source can be made to look very much like that of an impulsive source. Vibrators also come in all shapes and sizes, from 20 ton trucks which can generate both compressional and shear energy, to highly portable, hand-held devices, generating roughly the power of a hi-fi loudspeaker. Sparkers, which rely on electric discharges, are used for near-surface surveys conducted before siting offshore oil platforms. Airguns use compressed air from the ship to generate pulses similar to that of an explosion.[1]

The time-series, or seismogram, recorded by each receiver group is called a trace. The set of traces recorded by all the receivers for a given source is called a **common source gather**. There are three other standard ways of organizing traces into gathers. We could look at all the traces recorded by a given receiver group for all the sources. This is called a **common receiver gather**. We could look at all the traces whose source-receiver distance is a fixed value (called the offset). This called **common offset gather**. Finally we could look at all the traces whose source-receiver midpoint is a fixed value. This is a **common midpoint gather**.

Figure 1.1 illustrates these coordinate systems for a two-dimensional (2d) marine survey. (It's called 2d since the goal is to image a vertical plane through the earth, even though the data were recorded along a line.) The source-receiver coordinates r and s are connected to the offset-midpoint coordinates x and h via a $\pi/4$ rotation:

$$x = (r+s)/2$$
$$h = (r-s)/2 \tag{1.1}$$

---

[1] For more details on sources and receivers see [39] and [17].

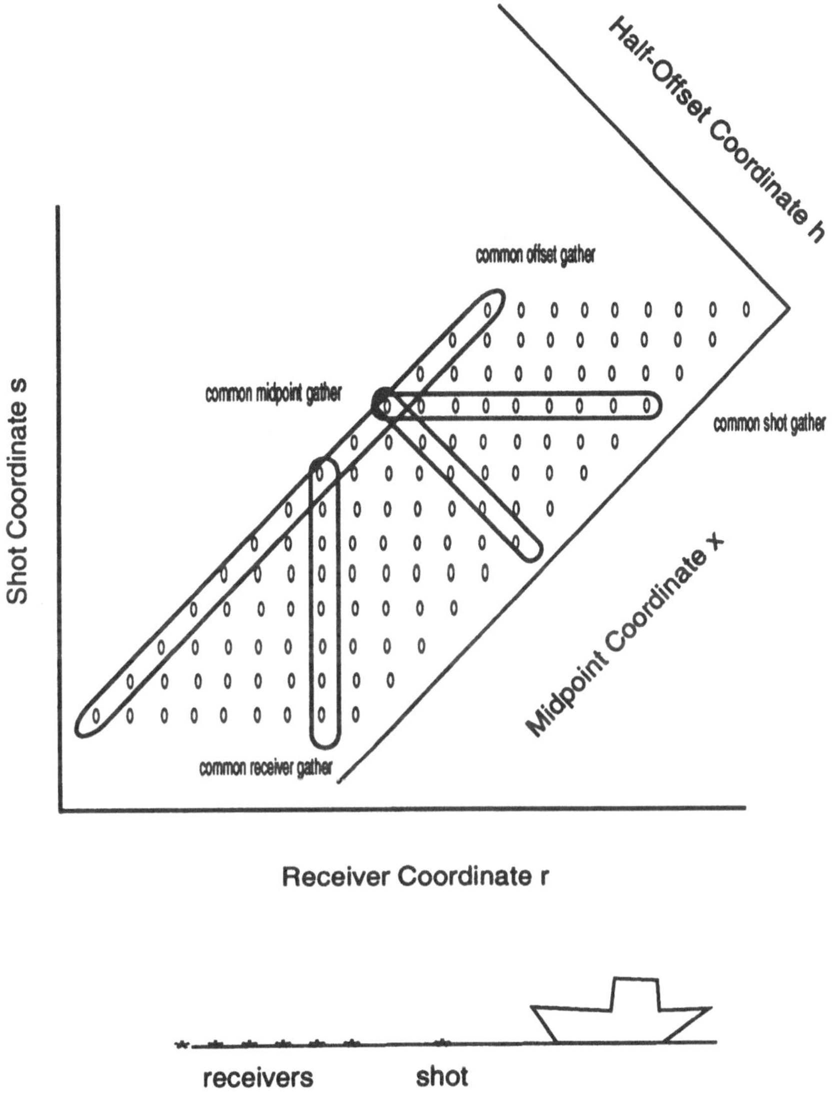

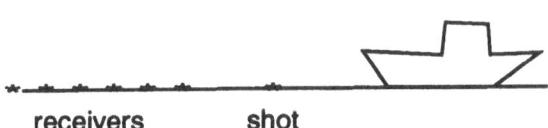

Fig. 1.1. Relationship among the coordinates r, s, h, and x. Each 0 indicates a seismic trace.

For the most part in this course we will be discussing post-stack, or zero-offset, migration methods, in which case the common midpoint domain (CMP) is generally used. The fold of a CMP gather is the number of source/receiver pairs for each midpoint. This is illustrated in Figure 1.2 which shows three-fold coverage. In this example the distance between sources is half the distance between receivers (Figure 1.3). Therefore the CMP spacing will be equal to the shot spacing and half the receiver spacing (known as the group interval). In Figure 1.3 the source/receiver raypaths $S_1 - R_5$ and $S_2 - R_4$ are associated with the same CMP location at R_2.

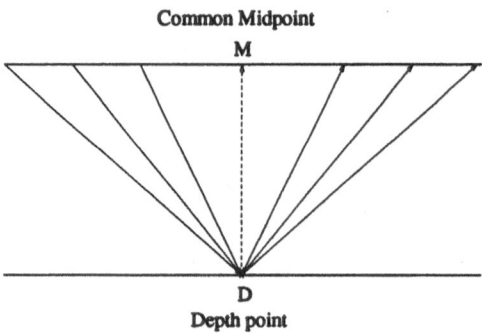

Fig. 1.2. Common midpoint ray paths under three-fold coverage. Sources are on the left and receivers are on the right.

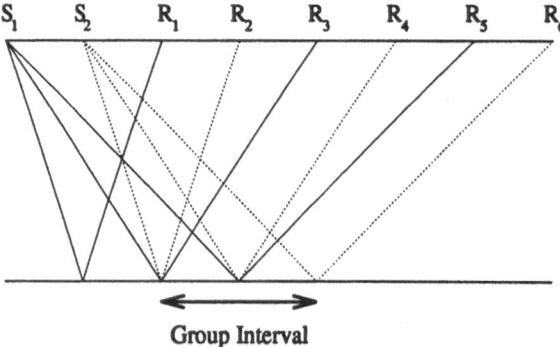

Fig. 1.3. In this example the shot spacing is half the group interval (the distance between receivers on a given line). Thus the CMP spacing is half the group interval too.

1.2 Huygens' Principle

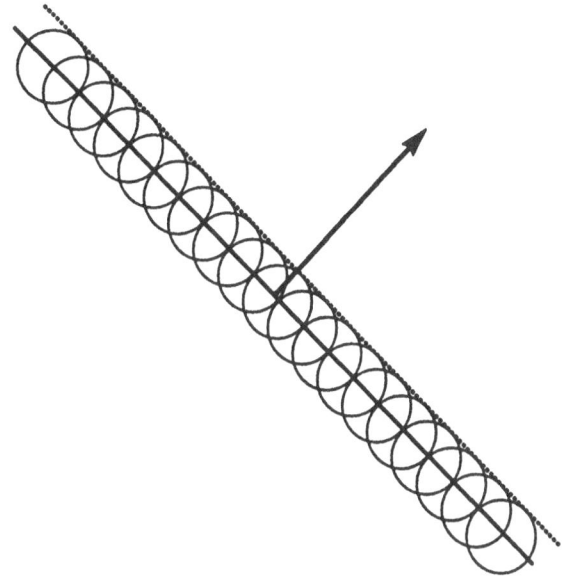

Fig. 1.4. Huygens' Principle: every point on a wavefront can be considered as a secondary source of spherical wavelets. Here the plane wave moves a distance $v\delta t$ in the time increment δt. Its new position is found by drawing a line through the envelope of all of the spherical wavelets.

Huygens' principle, that every point on a wavefront (propagating in an isotropic medium) can be considered a secondary source of spherical wavelets, allows us to reduce problems of wave propagation and de-propagation (migration) to the consideration of point sources. Figure 1.4 shows Huygens' construction for a plane wave propagating to the upper right. Each secondary spherical wavelet satisfies the equation

$$(z - z_0)^2 + (y - y_0)^2 + (x - x_0)^2 = v^2(t - t_0)^2 \tag{1.2}$$

where (x_0, z_0) is the origin of the spherical wavelet and t_0 is its initiation time. If the medium is homogeneous, we can use spherical wavelets of finite radius, otherwise the construction must proceed in infinitesimal steps.

If we could view snapshots of the Earth's subsurface after a point source were initiated, we would see an expanding spherical wavelet just as when a stone is thrown into a pond. So Equation (1.2) defines a family of circles in $x - z$ space (image space). But that's not how we actually see the seismic response of a point source. We record data for a range of times at a fixed depth position z (usually $z = 0$). If z is fixed, Equation (1.2) defines a family of hyperbolae in $x - t$ space (data space). So just as the spherical wavelet is the fundamental unit out of which all other wave propagation phenomena can be built (via Huygens' principle), so the hyperbola is the fundamental feature of reflection seismology, in terms of which all other reflection/diffraction events recorded at the surface can be interpreted.

To be precise, what we really need is a generalization of Huygens' principle due to Fresnel, who combined the wavelet construction with the Young's principle of interference. In other words, the expanding spherical wavelets are allowed to interfere with one another. Further, you will notice in Figure 1.4 that the spherical wavelets have two envelopes, one propagating in the same direction as the original plane wave and one propagating in the opposite direction. We only want the first one of these. Fresnel solved this problem by introducing an angle-dependent scaling factor called the "obliquity" factor. We will see later when we discuss integral methods of migration how this obliquity factor arises automatically.

Fresnel was the first person to solve problems of diffraction, which is any deviation from rectilinear propagation other than that due to reflection and refraction.[2] Fresnel's prediction that there should be a bright spot in the geometrical shadow of a disk was a major blow to the corpuscular theory of light. In fact, this surprising prediction was regarded as a refutation of his memoir by French Academician Poisson. Fortunately, another member of the committee reviewing the work, Arago, actually performed the experiment and vindicated Fresnel [8].

1.3 Zero-Offset Data

So with the Huygens-Fresnel principle in hand, we can think of any reflection event as the summation of the effects of a collection of point scatterers distributed along the reflecting surface. Each one of these point scatterers is responsible for a diffraction hyperbola. In order to be able to make a picture of the reflector, which is the goal of migration, we need to be able to collapse these diffraction hyperbolae back to their points of origin on the reflector. There is one circumstance in which

---

[2] Geometrical optics is the limit of wave optics as the wavelength λ goes to zero. In this limit there is no diffraction: there is no light in the shadow of a sharply defined object. So in contrast to refraction which affects the shorter wavelengths more strongly, diffraction tends to deflect the "red" end of the spectrum more than the "blue" end; the deflection being relative to the geometrical optics ray path. The coronae around the sun and moon are diffraction effects caused by the presence of water droplets randomly distributed in a layer of haze and are strongest when the droplets are of approximately uniform size, whereas the halos around the sun and moon are caused by refraction through ice crystals in thin cirrus clouds (See [43], Chapter V).

this procedure is especially easy to visualize: zero-offset seismograms recorded in a constant velocity earth. Imagine that we can place a seismic source and receiver right next to one another; so close in fact they are effectively at the same location. With this arrangement we can record the seismic echos from the same point whence they originated. (To be precise, if we start recording at the same time that the source goes off we will record the direct arrival wave, and no doubt shake up the receiver pretty thoroughly too. Therefore in practice we must mute the direct arrival; either that or wait until just after things have settled down to turn on the receiver.) Then we pick up our apparatus and move down the survey line a bit and repeat the procedure. This is called zero offset experiment. In practice we can approximate the zero-offset section by selecting the common-offset section having the minimum offset.

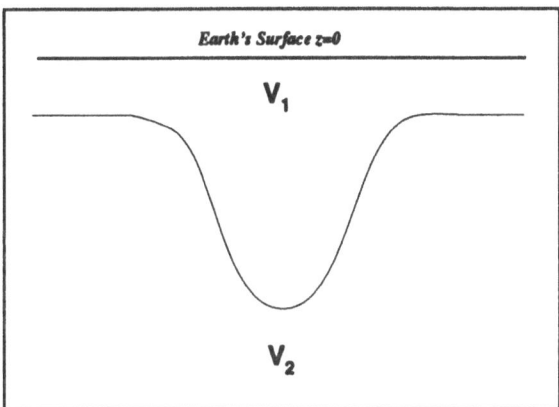

Fig. 1.5. Simple model of a syncline. The velocity above and below the syncline is assumed to be constant.

Figure 1.5 shows a model of a geologic syncline. The velocity is assumed to have the constant value v_1 in the layer above the syncline, and a different value v_2 in the layer below. This jump in the velocity gives rise to a reflection coefficient which causes the downgoing pulse to be partially reflected back to the surface where it is recorded on the geophone.

Figure 1.6 is a computer generated zero-offset section collected over this model. Notice that the travel time curve becomes multi-valued as the source receiver combiniation moves over the syncline. That's because there is more than one zero-offset ray arriving at the receiver in the time window we have recorded. **The goal of migration is to make Figure 1.6 look as much like Figure 1.5 as possible.**

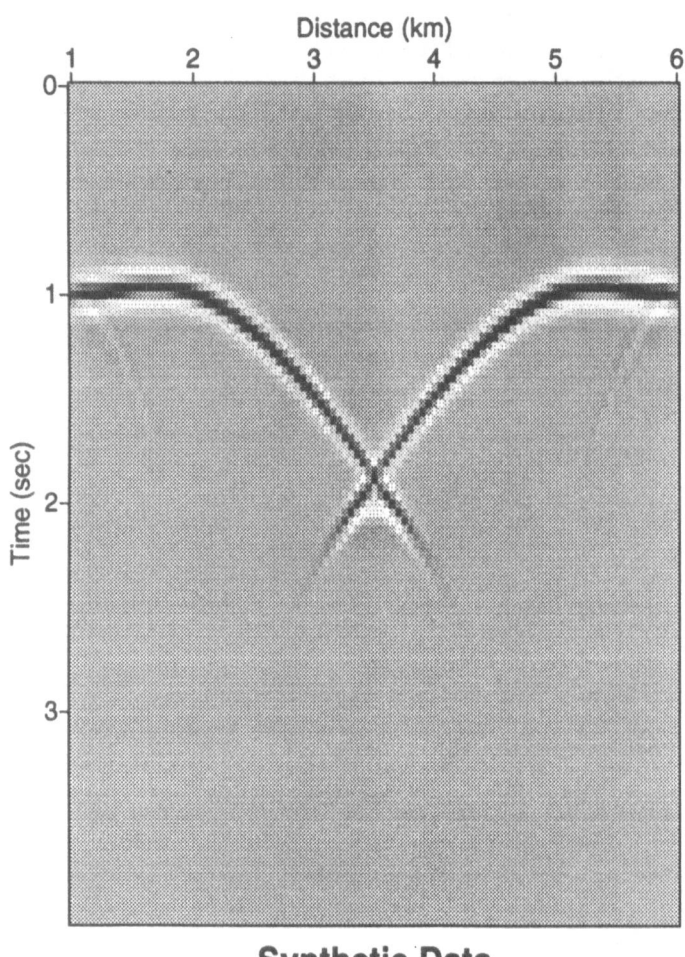

Synthetic Data

Fig. 1.6. Synthetic zero-offset data recorded over the syncline model shown in Figure 1.5. Shown are 100 traces sampled at .04 seconds per sample, spanning 6 km. on the Earth's surface.

1.4 Exploding Reflector Model

If T is the two-way time for a signal emanating at $t = 0$ to travel from the source to the reflector and back to the surface, then it follows that $T/2$ is the time it takes for a signal to travel from the source down to the reflector, or from the reflector back up to the source. So apart from (let's assume for the moment minor) distortions, the data recorded in the zero-offset experiment are the same that would be recorded if we placed sources along the syncline and fired them off at $T/2$ with a strength proportional to the reflection coefficient at the syncline. Equivalently we could fire off the exploding reflectors at $t = 0$ and halve the velocity.

This means that if we somehow run the zero-offset wavefield recorded at the surface $p(x, y, z = 0, t)$[3] backwards in time, then by evaluating the results at $t = 0$ we would have a picture of the reflection event as it occurred $p(x, y, z, t = 0)$: provided we use half the true velocity when we run the recorded field backwards in time. As we will see in detail later on, in the case of a constant velocity medium this time shifting of the wave field can be accomplished by a simple phase shift, i.e., multiplying the Fourier transform of the recorded data by a complex exponential. For now, Figure 1.7 must suffice to show the results of just such a procedure. This is an example of phase-shift migration.

You will notice that the vertical axis is time not depth. This is referred to as migrated time or vertical travel time. In the simple case of constant velocity media, there is a simple scaling relationship between depth and time: $t = z/v$. This extends to depth-dependent or $v(z)$ media via

$$t = \int_0^z \frac{dz}{v(z)}. \tag{1.3}$$

When we use this migrated time rather than depth for the output image, the migration is called time migration. Evidently time migration will be a reasonable choice only if the velocity shows no significant lateral, or $v(x)$, variation. Time migration has the advantage that the migrated section is plotted in the same units as the input seismic section; it has the disadvantage of being less accurate than depth migration. The more the velocity varies laterally, the less accurate time migration will be.

---

[3] Invariably z is taken to be the vertical coordinate. But in earth science z increases downward.

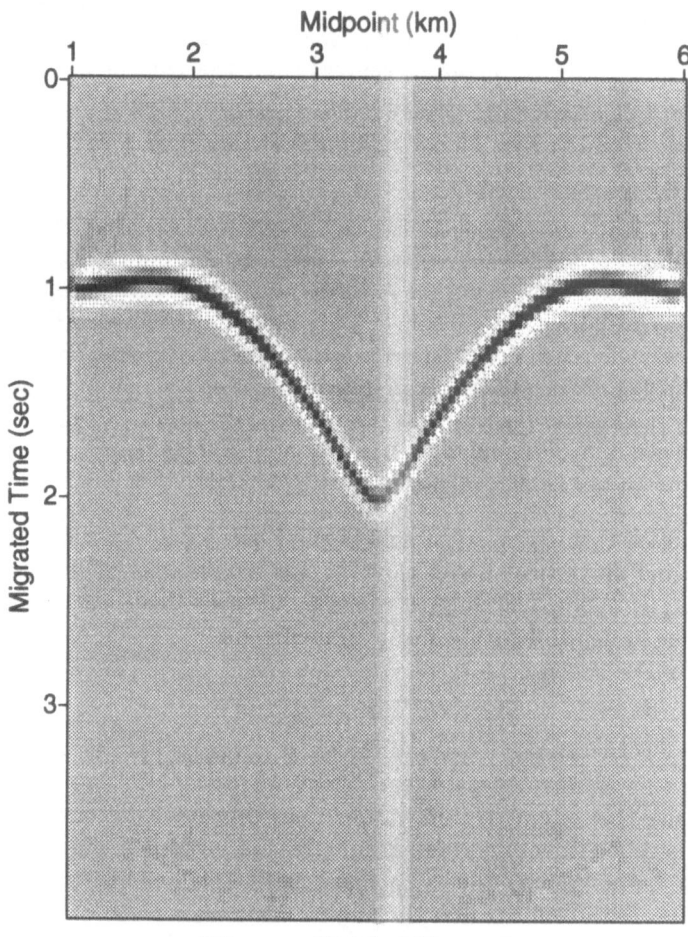

Phase Shift Migration

Fig. 1.7. Constant velocity phase shift migration using the correct velocity.

1.5 Finite Offset

The exploding reflector analogy is a powerful one. Unfortunately it only applies to zero-offset data, and typically exploration seismic surveys involve kilometers of offset between each source and the farthest receivers. Further, there are even true zero-offset situations where the exploding reflector concept fails, such as in the presence of strong lenses or when rays are multiply reflected.

Life gets more complicated when there is offset between the sources and receivers. Instead of there being a single two-way travel time for each source, we have a travel time curve as a function of offset. Figure 1.8 shows a CMP gather associated with 6 source/receiver pairs. The dotted curve is the hyperbolic travel time versus offset curve for a constant velocity layer.

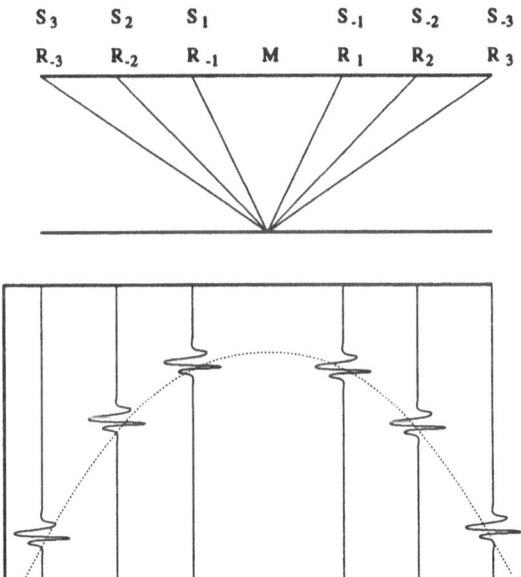

Fig. 1.8. Hyperbolic travel time curve associated with a CMP gather over a flat, constant velocity layer.

If the depth of the reflecting layer is z, and we let h denote half the distance between source and receiver, then by the Pythagorean theorem it follows that the two-way travel time as a function of half-offset h is given by

$$t(h)^2 \left(\frac{v}{2}\right)^2 = h^2 + z^2 \tag{1.4}$$

where v is the (constant) velocity in the layer above the reflector.

Now the zero-offset two-way travel time is simply

$$t_0 = 2\frac{z}{v}. \tag{1.5}$$

So, the difference between the two is

$$t(h)^2 - t_0^2 = \left(\frac{2h}{v}\right)^2. \tag{1.6}$$

This is called the normal moveout correction and is the amount by which the real traces have to be shifted in order to make them approximate the results of a zero-offset experiment. Normal moveout (NMO) refers generally to the hyperbolic moveout associated with flat layers. In the event that we have a stack of flat layers, the travel time curves are still approximately hyperbolic. If we apply the NMO correction to traces sharing a common midpoint and sum them, we get a plausible approximation to a zero-offset trace. This means that we can apply zero-offset migration to seismic data with nonzero offset. Further, stacking the traces in this way results in significant improvement in signal to noise; not only is random noise attenuated by stacking, but nonhyperbolic events are attenuated as well. Stacking is often the only way to see weak, noise-contaminated events. And the resulting post-stack migration results are sometimes surprisingly good even in cases where common midpoint (CMP) stacking should be a poor approximation to the zero-offset trace.

Further, if we knew that the reflecting layer were horizontal, then we could estimate the velocity above it by measuring the slope of the asymptote to the hyperbola. This is illustrated in Figure 1.9. The slope of the hyperbolic asymptote is just one over the velocity in the layer above the reflector. And the hyperbola's apex is at the one-way zero-offset travel time.

1.6 Stacking Velocity Analysis

In order to correct for normal moveout prior to stacking we must have a velocity model (v in Equation (1.6)). This is equivalent to stacking along the right hyperbola as shown in Figure 1.10. If we choose too high or too low a velocity then stacking will not produce a very high power in the output trace. We could imagine doing a suite of stacks associated with different velocities and zero-offset travel times and plotting the resulting stack power. At each zero-offset travel time (equivalent to depths in a constant v or $v(z)$ medium) we could scan across velocity and pick the v which gives the highest stack power. Proceeding down in depth

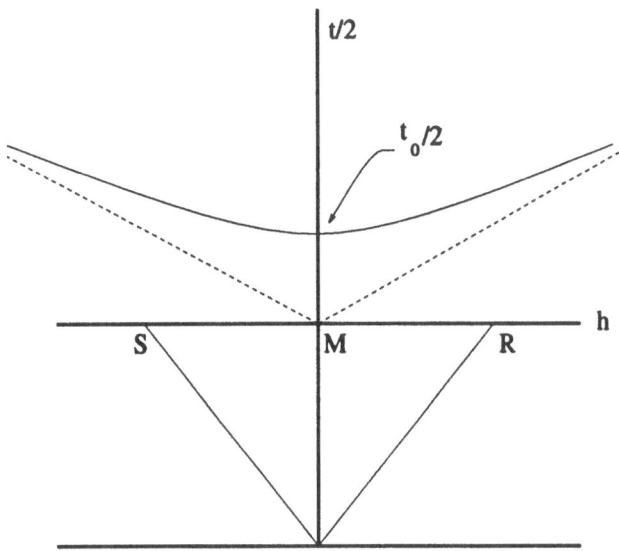

Fig. 1.9. The travel time curve for a constant velocity, flat-layer model is a hyperbola whose apex is at the one-way zero-offset travel time $t_0/2$. After [39].

or zero-offset travel time in this way we could map out the depth dependence of stacking velocity automatically.

We can illustrate this "stacking velocity analysis" with a simple example. Consider an Earth model which is composed of 5 flat, horizontal layers. Figure 1.11 shows 10 common source gathers for such a model assuming a constant velocity of 2 km/second. The stacking velocity panel for this is shown on the right in Figure 1.11. The sweep over velocity consisted of 50 increments of 50 m/s. The stacking panel shows maximum coherency at a constant, or depth-independent, velocity, but the coherence maxima are somewhat smeared out. Next, if we put in a depth-dependent velocity ($dv/dx = 1$), then the common source gathers look like the left side of Figure 1.12. The stacking velocity panel is shown on the right. It's clear that we would have no difficulty picking out the linear trend in velocity.

1.7 Dipping Reflectors

So far we have only considered the reflection seismology of flat (i.e., horizontal) layers. But because of tectonic and other forces (such as buoyant material welling up under the influence of gravity), dipping layers are very common–in fact mapping apparent dip (that which we see on the zero-offset seismic sections) into true dip is one of the principle goals of migration. Unless the dip is actually zero,

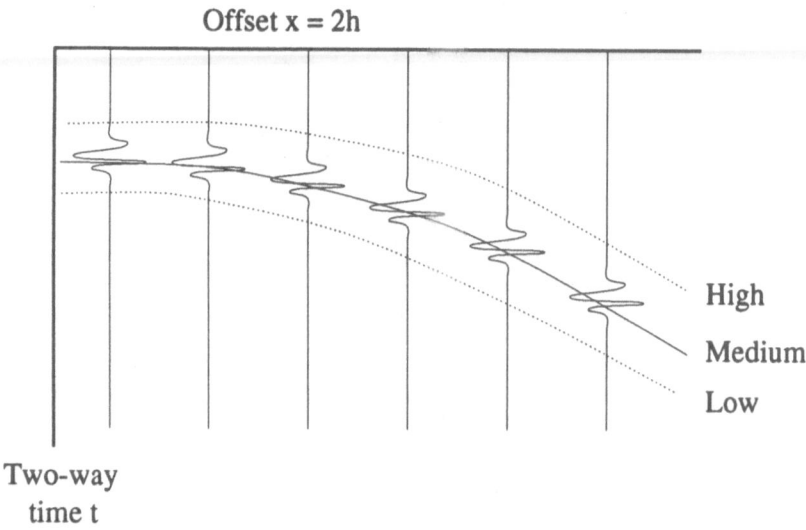

Fig. 1.10. Stacking along successive hyperbolae as we vary the velocity and zero-offset travel time gives a stack power $S(v, t_0)$ as a function of v and t_0. Finding the "best" velocity model is then a matter of optimizing the stack power function.

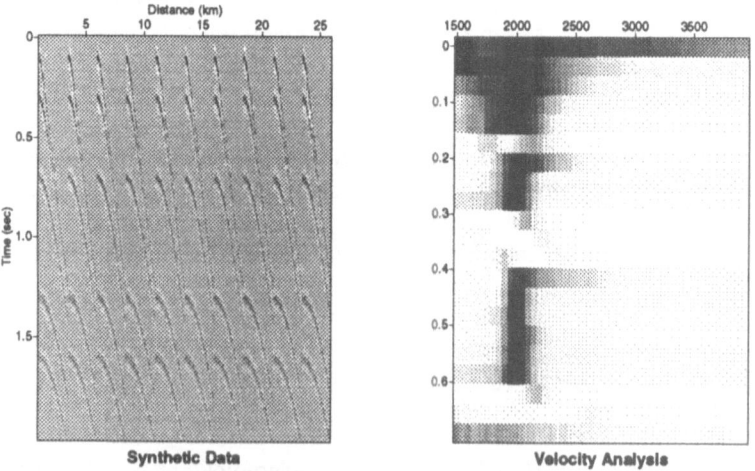

Fig. 1.11. 10 common source gathers for a 5 layer constant velocity model (left). Stacking velocity analysis for these gathers (right).

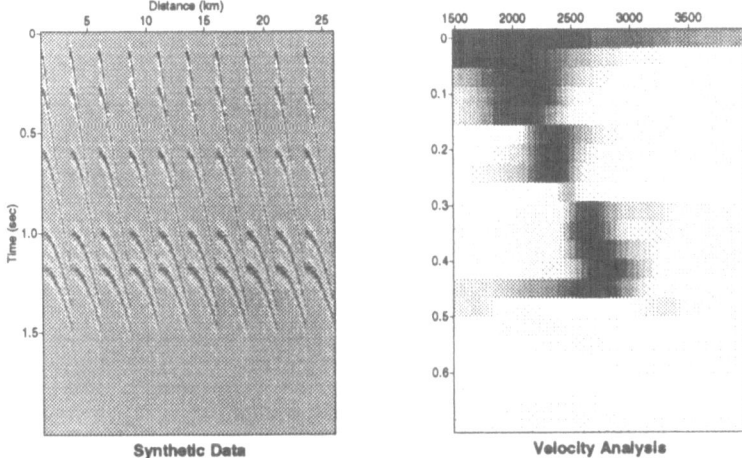

Fig. 1.12. 10 common source gathers for a 5 layer model with a constant $v(z)$ gradient (left). Stacking velocity analysis for these gathers (right).

true dip is always greater than apparent dip. Consider the limiting case of reflections from a vertical layer (90 degree dip). Since the reflections must propagate along the surface, the zero-offset section will have a linear moveout whose slope is inversely proportional to the velocity.

More generally consider a reflector dipping at an angle of θ in the true earth (cf. Figure 1.13). The relation between depth and offset is: $z = x \tan \theta$. Now the normal incidence (i.e., zero-offset) travel time for a ray propagating from x down to the reflector and back up again is $t = 2r/v$, where r is the ray path length. But $r = x \sin \theta$. So that we may compare apparent and true dip, we need to convert this travel time to depth via $z = vt/2$. Therefore in the unmigrated depth section we have: $z = x \sin \theta$. The slope of this event is, by definition, the tangent of the apparent dip angle. Call this angle β. Therefore we have $\tan \beta = \sin \theta$. This is called the migrator's equation and shows that apparent dip is always less than true dip.

Let's carry this analysis one step further and consider a segment of dipping reflector as in Figure 1.14. The events associated with the two normal incidence rays drawn from the dipping reflector to the two receivers will appear on the unmigrated section at the offset locations associated with the two receivers. Therefore not only will migration increase the apparent dip of this segment of the reflector, but it will have to move the energy horizontally (updip) as well.

To recap what we have shown thus far in this introductory chapter, migration will:

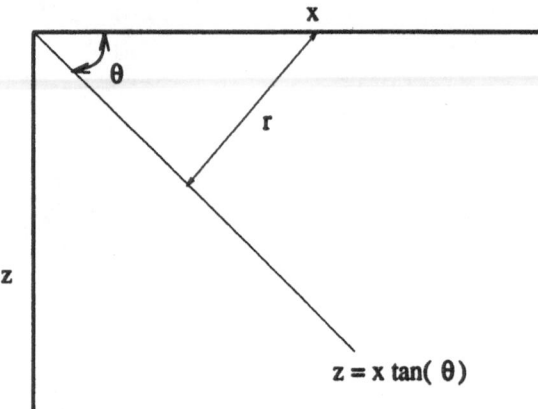

Fig. 1.13. Unless the reflector is horizontal, the apparent dip on the time section is always less than the true dip in the earth.

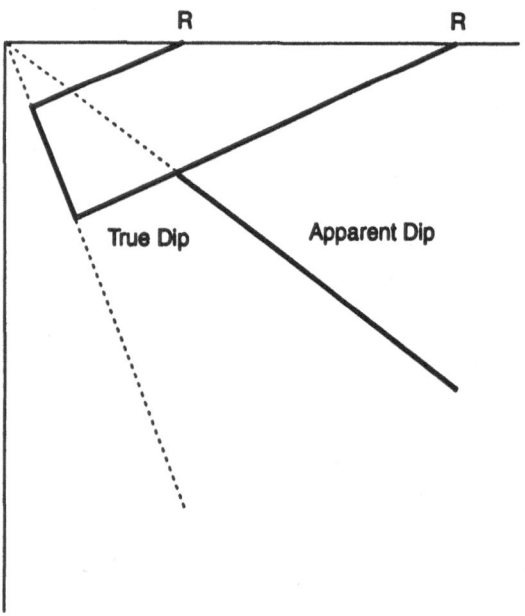

Fig. 1.14. In addition to increasing apparent dip, migration also moves energy updip.

 — Focus energy by collapsing diffraction hyperbolae.

 — Increase apparent dip.

 — Move energy horizontally in the updip direction.

1.8 Resolution

Just as the stacking velocity panels in the last section did not collapse to points, so diffraction hyperbolae do not collapse to points when migrated, they focus. The spread of the focus tells us something about the resolution of our imaging methods. Vertical resolution is governed by the wavelength of the probing wave: v/f, where f is the frequency. But we need to remember to divide v by two for the exploding reflector calculation; we divide in two again to get the length of one half cycle. So the effective vertical resolution of a 50 hz seismic wave propagating in a medium whose wavespeed is 3000 m/s is about 15 m [14].

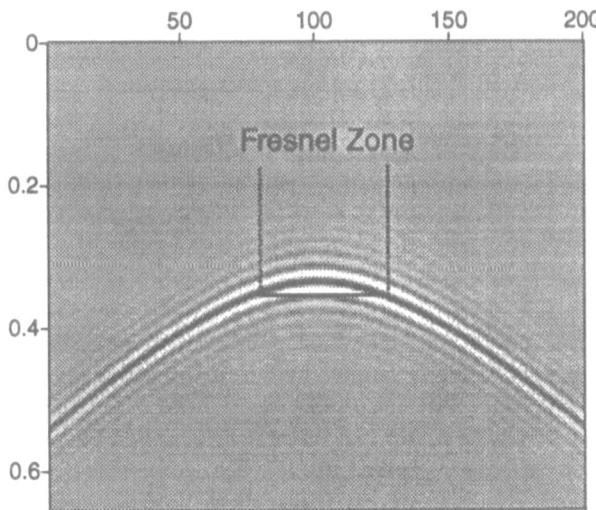

Fig. 1.15. The first Fresnel zone is the horizontal distance spanned by one half a wave-length.

Horizontal resolution is usually measured in terms of the Fresnel zone. Imagine a spherical wave striking the center of a hole punched in an infinite plate. How wide would the hole have to be to allow the sphere to penetrate to one half of its wavelength λ? Answer: by definition, one Fresnel zone. Alternatively we can measure the Fresnel zone across a diffraction hyperbola as the distance spanning one zero crossing (Figure 1.15). This concept goes back to the Huygens-Fresnel principle previously discussed. Picture a point source in an infinite homogeneous medium. As we will see later, the spatial part of the solution of the wave equation in this case is e^{ikr}/r where r is the distance from the origin of the disturbance to the wavefront and k is the reciprocal wavelength. Figure 1.16 shows the spherical wavelet at some point in time after its origination at the point P. To get the field at some observation point Q we consider each point on the spherical wavefront as a secondary source of radiation. In other words, the seismogram at point Q, $S(Q)$ must be proportional to the integral over the spherical wavefront of

$$\frac{e^{ikr}}{r} \int_\sigma \frac{e^{iks}}{s} K(\chi) d\sigma \qquad (1.7)$$

where σ refers to the surface of the spherical wavefront, χ is the angle between the position vector and the propagation direction, and K is an angle-dependent fudge factor that Fresnel had to add to make the wave have the proper directional dependence.[4] The first exponential, the one outside the integral, is just the point-source propagation from P to the spherical wavefront. The second exponential, inside the integral, is the propagation from the wavefront to the observation point Q.

The key idea for this discussion is how Fresnel actually computed the integral. As illustrated in Figure 1.16, he broke up the sphere into zones whose radii (measured from the observation point Q) differed by one-half wavelength $\lambda/2$. The total radiation at the observation point is then just the sum of the contributions from the individual zones. Ths surprising conclusion is: the total disturbance at Q is equal to half the disturbance due to the first zone alone. And since the intensity is the square of the disturbance, if we could somehow screen off all but the first zone, the intensity recorded at Q would be 4 times as large as if there were no screen there. Claerbout ([14], page 19) gives an analogy of shouting at a hole in the Berlin wall and says that holes larger than one Fresnel zone cause little attenuation. This might be true for an irregularly shaped hole, but the disturbances due to the even and odd numbered zones are of opposite signs. In fact, it can be shown that the disturbances due to the first two zones are nearly equal. Therefore, if all but the first two zones are somehow covered up, the disturbance at the observation point will be nearly zero! This is worth repeating: if we go from having all but the first zone screened off to having all but the first two, the intensity measured at Q will go from 4 times the unobstructed intensity to zero. For more details see [8] (Chapter VIII) and [43] (Chapter V). Especially note Section 35 of [43], which discusses diffraction from sharp obstacles. In this section Sommerfeld reproduces an astonishing "photograph" taken by E. von Angerer using a circular sheet-metal disk!

---

[4] This obliquity factor will be explained rigorously later when we get to Kirchhoff migration.

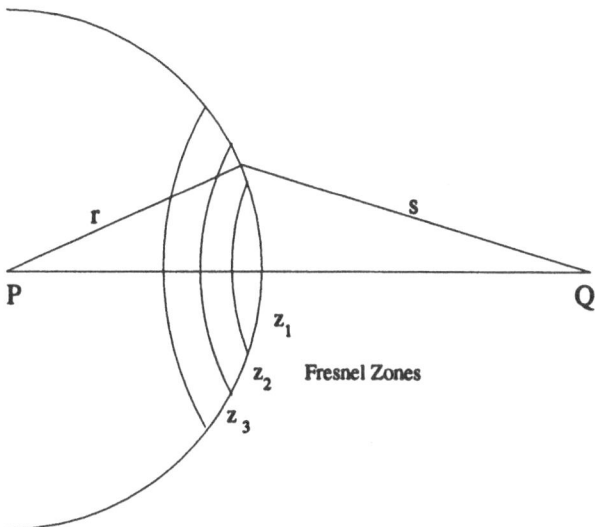

Fig. 1.16. Fresnel zone construction

Exercises

1.1 What is the vertical travel time for a medium whose velocity increases linearly with depth?

1.2 What are the angles of incidence and reflection for zero-offset rays? Does your answer depend on the reflector geometry?

1.3 The travel time curve for a common source record in a model with a single, flat constant-velocity layer is a hyperbola. What is it if there are two flat layers instead? What is the NMO correction in this case?

1.4 A point scatterer in image space gives rise to a hyperbolic event in data space. What reflector geometry would give rise to a point in data space?

1.5 What are the minimum and maximum vertical resolution in a 5 layer model whose wavespeeds are 2.0, 3.0, 3.5, 4.5, and 5.0 km/second?

1.6 Look at Figure 1.11 again. Why are the hyperbolae associated with the deeper reflectors flatter looking than the ones associated with the shallower reflectors?

1.7 Figure 1.13 shows the geometry for a reflector with arbitrary dip θ. The velocity above the reflecting layer is a constant v. Consider the zero-offset section recorded over such a reflector. Convert the curve of zero-offset travel times versus offset to migrated times. Let the dip of the resulting curve be called

β. What is the relationship between θ and β? This is called the migrator's equation.

1.8 Show that apparent dip is always less than migrated dip.

1.9 Computer Exercise: I

In most of the computer exercises in this course we will take advantage of the SU (Seismic Unix) processing system developed at the Center for Wave Phenomena. Instructions on how to obtain SU, as well as a User's Guide, are in Appendix A.

To get a list of all the executables (including graphics) type suhelp. Then simply type the name of any executable (with no command line arguments) to get information on how to run the program.

For example, here is what happens if you type sufilter at the Unix prompt:

```
SUFILTER - applies a zero-phase, sine-squared tapered filter

sufilter <stdin >stdout [optional parameters]

Required parameters:
        if dt is not set in header, then dt is mandatory

Optional parameters:
    f=f1,f2,...                 array of filter frequencies(HZ)
    amps=a1,a2,...              array of filter amplitudes
    dt = (from header)          time sampling rate (sec)

Defaults: f=.10*(nyquist),.15*(nyquist),.45*(nyquist),
.50*(nyquist) (nyquist calculated internally)
        amps=0.,1.,...,1.,0.   trapezoid-like bandpass
filter

Examples of filters:
Bandpass:
    sufilter <data f=10,20,40,50 | ...
Bandreject:
    sufilter <data f=10,20,30,40 amps=1.,0.,0.,1. | ..
Lowpass:
    sufilter <data f=10,20,40,50 amps=1.,1.,0.,0. | ...
Highpass:
    sufilter <data f=10,20,40,50 amps=0.,0.,1.,1. | ...
```

```
Notch:
    sufilter <data f=10,12.5,35,50,60 \
    amps=1.,.5,0.,.5,1. |..
```

Here is an example shell script for generating some synthetic data and then migrating it.

```
n1=101
n2=101
#
## use susynlv to make synthetic data
susynlv nt=$n1 dt=0.04 ft=0.0 nxo=1 \
   nxm=$n2 dxm=.05 fxm=0.0 er=0 ob=1  \
   v00=1.0 dvdz=0 dvdx=0 smooth=1 \
   ref="0,.5;1.0,.5;2.,.8;2.5,1.0;3.0,.8;4.0,.5;5.0,.5"  |
sushw key=d2 a=.05 > junk.susyn

supsimage  < junk.susyn label1="Time (sec)" \
label2="Distance (km)" \
          title="Synthetic Data" > synthetic_data.ps

# apply gazdag
sugazmig  < junk.susyn tmig=0 vmig=1  > junk.out
supsimage < junk.out label1="Migrated Time (sec)" \
label2="Midpoint (km)" \
          title="Phase Shift Migration"  > migrated_data.ps
```

In this example we've used the postscript based plotting routine supsimage. The file migrated_data.ps, for example, can be displayed on an X-windows device using ghostscript (gs) or sent to a laser printer. See the information in suhelp under plotting for more details.

Exercises

1.9 Dip: Use susynlv to make 100 zero-offset traces over a single flat (horizontal) layer 1km deep. Make the model 5 km across and use a constant velocity of 2 km/sec, use an offset (or trace) spacing of 50 m. and a sample interval of 8 ms. Then redo this experiment, varying the dip of the layer from 0 to 60 degrees in 20 degree steps. So you should end up with 4 zero-offset sections. Measure the apparent dip of each section.

1.10 Offset: For the same 4 models as in the first problem, record one common source gather of 100 traces with a maximum offset of 5km.

1.11 Band limited data: Take one of your zero-offset sections from above and use sufilter to band limit it to a reasonable exploration-seismic range of 5-50 hz.

1.12 Migration: Use susynlv to generate a zero-offset section over a layer dipping at 45 degrees. Use a constant velocity of 2 km/sec and record 100 traces spaced 50 m apart. Migrate your zero-offset section with sugazmig and verify that migration increases apparent dip.

2. Harmonic Analysis, Delta Functions, and All That

We collect here a few miscellaneous but essential results results about Fourier Transforms, Fourier Series, the sampling theorem, and delta functions.

2.1 Fourier Series

Suppose f is a piecewise continuous function periodic on the interval $[0, 2\pi]$. Then the Fourier coefficients of f are defined to be

$$c_n = \frac{1}{2\pi} \int_0^{2\pi} f(t)e^{int}\, dt \qquad (2.1)$$

The Fourier series for f is then

$$f(t) = \sum_{n=-\infty}^{\infty} c_n e^{-int}. \qquad (2.2)$$

One has to be a little careful about saying that a particular function is equal to its Fourier series since there exist piecewise continuous functions whose Fourier series diverge everywhere! However, here are two basic results about the convergence of such series.

Pointwise Convergence Theorem: If f is piecewise continuous and has left and right derivatives at a point c[1] then the Fourier series for f converges converges to

$$\frac{1}{2}\left(f(c-) + f(c+)\right) \qquad (2.3)$$

---

[1] A right derivative would be: $\lim_{t \to 0}(f(c+t) - f(c))/t$, $t > 0$. Similarly for a left derivative.

where the + and - denote the limits when approached from greater than or less than c.

Another basic result is the **Uniform Convergence Theorem**: If f is continuous with period 2π and f' is piecewise continuous, then the Fourier series for f converges uniformly to f. For more details, consult a book on analysis such as *The Elements of Real Analysis* by Bartle [3] or *Real Analysis* by Haaser and Sullivan [27].

2.2 Fourier Transforms

If a function is periodic on any interval, that interval can be mapped into $[-\pi, \pi]$ by a linear transformation. Further, if a function is not periodic, but defined only on a finite interval, it can be replicated over and over again to simulate a periodic function. (This is called periodic extension.) In all of these cases, the Fourier series applies. But for functions which are not periodic and which are defined over an infinite interval, then we must use a different argument. In effect, we must consider the case of a nonperiodic function on a finite interval, say $[-r, r]$, and take the limit as $r \to \infty$. If we do this, the Fourier series becomes an integral (cf. [27], Chapter 11, section 11). The result is that a function $f(t)$ is related to its Fourier transform $f(\omega)$ via:

$$f(t) = \frac{1}{2\pi} \int_{-\infty}^{\infty} f(\omega)e^{-i\omega t} \, d\omega \tag{2.4}$$

and

$$f(\omega) = \int_{-\infty}^{\infty} f(t)e^{i\omega t} \, dt \tag{2.5}$$

Here, using time and frequency as variables, we are thinking in terms of time series, but we could just as well use a distance coordinate such as x and a wavenumber k:

$$f(x) = \frac{1}{2\pi} \int_{-\infty}^{\infty} f(k)e^{-ikx} \, dk \tag{2.6}$$

with the inverse transformation being

$$f(k) = \int_{-\infty}^{\infty} f(x)e^{ikx} \, dx. \tag{2.7}$$

It doesn't matter how we split up the 2π normalization. For example, in the interest of symmetry many people define both the forward and inverse transform with a $1/\sqrt{2\pi}$ out front. It doesn't matter as long as we're consistent. We could get rid of the normalization altogether if we stop using circular frequencies ω in favor of f measured in hertz or cycles per second. Then we have

$$g(t) = \int_{-\infty}^{\infty} g(f)e^{-2\pi i f t} \, df \tag{2.8}$$

and

$$g(f) = \int_{-\infty}^{\infty} g(t)e^{2\pi i f t} \, dt \tag{2.9}$$

These transformations from time to frequency or space to wavenumber are invertible in the sense that if we apply one after the other we recover the original function. To see this plug Equation (2.7) into Equation (2.6):

$$f(x) = \frac{1}{2\pi} \int_{-\infty}^{\infty} dk \int_{-\infty}^{\infty} f(x')e^{-ik(x'-x)} \, dx'. \tag{2.10}$$

If we define the kernel function $K(x - x', \mu)$ such that

$$K(x' - x, \mu) = \frac{1}{2\pi} \int_{-\mu}^{\mu} e^{-ik(x'-x)} \, dk = \frac{\sin \mu(x'-x)}{\pi(x'-x)} \tag{2.11}$$

then we have

$$f(x) = \int_{-\infty}^{\infty} f(x')K(x' - x)dx' \tag{2.12}$$

where $K(x' - x)$ is the limit (assuming that it exists)[2] of $K(x' - x, \mu)$ as $\mu \to \infty$. In order for this to be true $K(x' - x)$ will have to turn out to be a Dirac delta function.

We won't attempt to prove that the kernel function converges to a delta function and hence that the Fourier transform is invertible; you can look it up in most books on analysis. But Figure 2.1 provides graphical evidence. We show plots of this kernel function for $x = 0$ and three different values of μ, 10, 100, 1000, and 10000. Clearly, in the limit that $\mu \to \infty$ the function K becomes a Dirac delta function.

---

[2] We're being intentionally fuzzy about the details to save time. If you've never encountered a careful treatment of "generalized functions" you should spend some time with a book like Volume I of Gel'fand and Shilov's *Generalized Functions* [24]. It's quite readable and very complete.

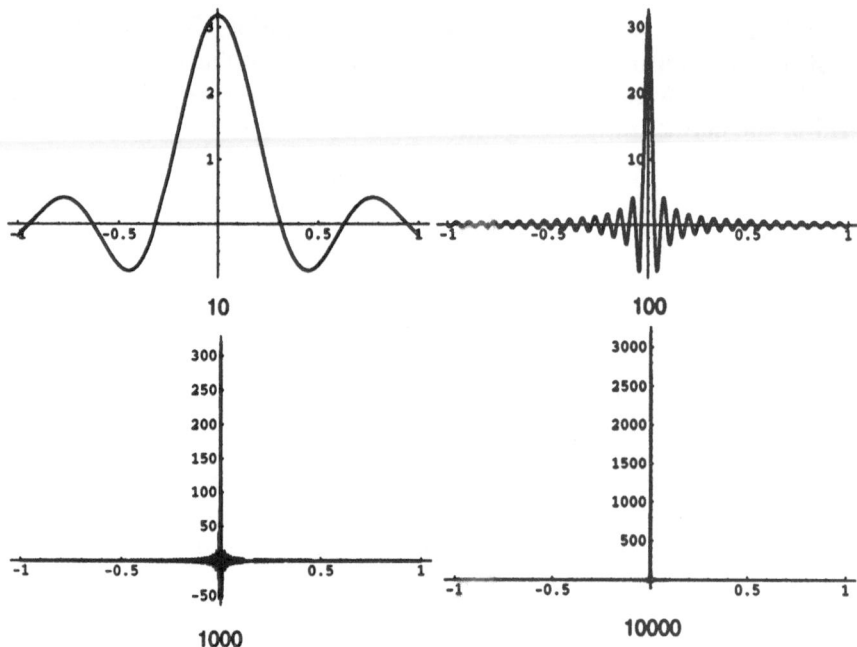

Fig. 2.1. The kernel $\sin \mu x / \pi x$ for $\mu = 10$, 100, 1000, and 10000.

2.2.1 Basic Properties of Delta Functions

Another representation of the delta function is in terms of Gaussian functions:

$$\delta(x) = \lim_{\mu \to \infty} \frac{\mu}{\sqrt{\pi}} e^{-\mu^2 x^2}. \tag{2.13}$$

You can verify for yourself that the area under any of the Gaussian curves associated with finite μ is one.

The spectrum of a delta function is completely flat since

$$\int_{-\infty}^{\infty} e^{-ikx} \delta(x) \, dx = 1. \tag{2.14}$$

For delta functions in higher dimensions we need to add an extra $1/2\pi$ normalization for each dimension. Thus

$$\delta(x, y, z) = \left(\frac{1}{2\pi}\right)^3 \int_{-\infty}^{\infty} \int_{-\infty}^{\infty} \int_{-\infty}^{\infty} e^{i(k_x x + k_y y + k_z z)} \, dk_x \, dk_y \, dk_z. \tag{2.15}$$

The other main properties of delta functions are the following:

$$\delta(x) = \delta(-x) \tag{2.16}$$

$$\delta(ax) = \frac{1}{|a|}\delta(x) \tag{2.17}$$

$$x\delta(x) = 0 \tag{2.18}$$

$$f(x)\delta(x-a) = f(a)\delta(x-a) \tag{2.19}$$

$$\int \delta(x-y)\delta(y-a)\,dy = \delta(x-a) \tag{2.20}$$

$$\int_{-\infty}^{\infty} \delta^{(m)}f(x)\,dx = (-1)^m f^{(m)}(0) \tag{2.21}$$

$$\int \delta'(x-y)\delta(y-a)\,dy = \delta'(x-a) \tag{2.22}$$

$$x\delta'(x) = -\delta(x) \tag{2.23}$$

$$\delta(x) = \frac{1}{2\pi}\int_{-\infty}^{\infty} e^{ikz}\,dk \tag{2.24}$$

$$\delta'(x) = \frac{i}{2\pi}\int_{-\infty}^{\infty} ke^{ikz}\,dk \tag{2.25}$$

2.3 The Sampling Theorem

Now returning to the Fourier transform, suppose the spectrum of our time series $f(t)$ is zero outside of some symmetric interval $[-2\pi f_c, 2\pi f_c]$ about the origin.[3] In other words, the signal does not contain any frequencies higher than f_c hertz. Such a function is said to be *band limited*; it contains frequencies only in the band $[-2\pi f_c, 2\pi f_c]$. Clearly a band limited function has a finite inverse Fourier transform

$$f(t) = \frac{1}{2\pi}\int_{-2\pi f_c}^{2\pi f_c} f(\omega)e^{-i\omega t}\,d\omega. \tag{2.26}$$

Since we are now dealing with a function on a finite interval we can represent it as a Fourier series:

$$f(\omega) = \sum_{n=-\infty}^{\infty} \phi_n e^{i\omega n/2f_c} \tag{2.27}$$

where the Fourier coefficients ϕ_n are to be determined by

---

[3] In fact the assumption that the interval is symmetric about the origin is made without loss of generality, since we can always introduce a change of variables which maps an arbitrary interval into a symmetric one centered on 0.

$$\phi_n = \frac{1}{4\pi f_c} \int_{-2\pi f_c}^{2\pi f_c} f(\omega)e^{-i\omega n/2f_c}\, d\omega. \tag{2.28}$$

Comparing this result with our previous work we can see that

$$\phi_n = \frac{f(n/2f_c)}{2f_c} \tag{2.29}$$

where $f(n/2f_c)$ are the samples of the original continuous time series $f(t)$. Putting all this together, one can show that the band limited function $f(t)$ is completely specified by its values at the countable set of points spaced $1/2f_c$ apart:

$$
\begin{aligned}
f(t) &= \frac{1}{4\pi f_c} \sum_{n=-\infty}^{\infty} f(n/2f_c) \int_{-2\pi f_c}^{2\pi f_c} e^{i(\omega n/2f_c - \omega t)}\, d\omega \\
&= \sum_{n=-\infty}^{\infty} f(n/2f_c) \frac{\sin(\pi(2f_c t - n))}{\pi(2f_c t - n)}.
\end{aligned} \tag{2.30}
$$

The last equation is known as the **Sampling Theorem**. It is worth repeating for emphasis: any band limited function is completely determined by its samples chosen $1/2f_c$ apart, where f_c is the maximum frequency contained in the signal. This means that in particular, a time series of finite duration (i.e., any real time series) is completely specified by a finite number of samples. It also means that in a sense, the information content of a band limited signal is infinitely smaller than that of a general continuous function.

So if our band-limited signal $f(t)$ has a maximum frequency of f_c hertz, and the length of the signal is T, then the total number of samples required to describe f is $2f_c T$.

2.3.1 Aliasing

As we have seen, if a time-dependent function contains frequencies up to f_c hertz, then discrete samples taken at an interval of $1/2f_c$ seconds completely determine the signal. Looked at from another point of view, for any sampling interval Δ, there is a special frequency (called the Nyquist frequency) given by $f_c = \frac{1}{2\Delta}$. The extrema (peaks and troughs) of a sinusoid of frequency f_c will lie exactly $1/2f_c$ apart. This is equivalent to saying that critical sampling of a sine wave is 2 samples per wavelength.

We can sample at a finer interval without introducing any error; the samples will be redundant, of course. However, if we sample at a coarser interval a very serious kind of error is introduced called aliasing. Figure 2.2 shows a cosine function sampled at an interval longer than $1/2f_c$; this sampling produces an apparent frequency of $1/3$ the true frequency. This means that any frequency component in the signal lying outside the interval $(-f_c, f_c)$ will be spuriously shifted into this interval. Aliasing is produced by undersampling the data: once that happens there is little that can be done to correct the problem. The way to prevent aliasing is to know the true band-width of the signal (or band-limit the signal by analog filtering) and then sample appropriately so as to give at least 2 samples per cycle at the highest frequency present.

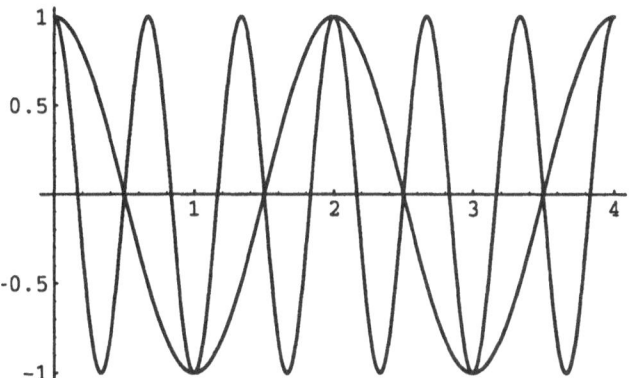

Fig. 2.2. A sinusoid sampled at less than the Nyquist frequency gives rise to spurious periodicities.

2.4 Discrete Fourier Transforms

In this section we will use the f (cycles per second) notation rather than the ω (radians per second), because there are slightly fewer factors of 2π floating around. You should be comfortable with both styles, but mind those 2πs! Also, up to now, we have avoided any special notation for the Fourier transform of a function, simply observing whether it was a function of space-time or wavenumber-frequency. Now that we are considering discrete transforms and real data, we need to make this distinction since we will generally have both the sampled data and its transform stored in arrays on the computer. So for this section we will follow the convention that if $h = h(t)$ then $H = H(f)$ is its Fourier transform.

We suppose that our data are samples of a function and that the samples are taken at equal intervals, so that we can write

$$h_k \equiv h(t_k), \qquad t_k \equiv k\Delta, \qquad k = 0, 1, 2, \ldots, N-1, \tag{2.31}$$

where N is an even number. In our case, the underlying function $h(t)$ is unknown; all we have are the digitally recorded seismograms. But in either case we can estimate the Fourier transform $H(f)$ at at most N discrete points chosen in the range $-f_c$ to f_c where f_c is the Nyquist frequency:

$$f_n \equiv \frac{n}{\Delta N}, \qquad n = \frac{-N}{2}, \ldots, \frac{N}{2}. \tag{2.32}$$

The two extreme values of frequency $f_{-N/2}$ and $f_{-N/2}$ are not independent ($f_{-N/2} = -f_{N/2}$), so there are actually only N independent frequencies specified above.

A sensible numerical approximation for the Fourier transform integral is thus:

$$H(f_n) = \int_{-\infty}^{\infty} h(t) e^{2\pi i f_n t}\, dt \approx \sum_{k=0}^{N-1} h_k e^{2\pi i f_n t_k} \Delta. \tag{2.33}$$

Therefore

$$H(f_n) \approx \Delta \sum_{k=0}^{N-1} h_k e^{2\pi i k n / N}. \tag{2.34}$$

Defining the *Discrete Fourier Transform* (DFT) by

$$H_n = \sum_{k=0}^{N-1} h_k e^{2\pi i k n / N} \tag{2.35}$$

we then have

$$H(f_n) \approx \Delta H_n \tag{2.36}$$

where f_n are given by Equation (2.32).

Now, the numbering convention implied by Equation (2.32) has $\pm$ Nyquist at the extreme ends of the range and zero frequency in the middle. However it is clear that the DFT is periodic with period N:

$$H_{-n} = H_{N-n}. \tag{2.37}$$

As a result, it is standard practice to let the index n in H_n vary from 0 to $N-1$, with n and k varying over the same range. In this convention 0 frequency occurs at $n = 0$; positive frequencies from from $1 \leq n \leq N/2 - 1$; negative frequencies run from $N/2 + 1 \leq n \leq N - 1$. Nyquist sits in the middle at $n = N/2$. The inverse transform is:

$$h_k = \frac{1}{N} \sum_{n=0}^{N-1} H_n e^{-2\pi i k n/N} \tag{2.38}$$

Mathematica, on the other hand, uses different conventions. It uses the symmetric normalization ($1/\sqrt{N}$ in front of both the forward and inverse transform), and defines arrays running from 1 to N in Fortran fashion. So in *Mathematica*, the forward and inverse transforms are, respectively:

$$H_n = \frac{1}{\sqrt{N}} \sum_{k=1}^{N} h_k e^{2\pi i(k-1)(n-1)/N} \tag{2.39}$$

and

$$h_k = \frac{1}{\sqrt{N}} \sum_{n=1}^{N} H_n e^{-2\pi i(k-1)(n-1)/N}. \tag{2.40}$$

If you are using canned software, make sure you know what conventions are being used.

2.4.1 Discrete Fourier Transform Examples

Here we show a few examples of the use of the DFT. What we will do is construct an unknown time series' DFT by hand and inverse transform to see what the resulting time series looks like. In all cases the time series h_k is 64 samples long. Figures 2.3 and 2.4 show the real (left) and imaginary (right) parts of six time series that resulted from inverse DFT'ing an array H_n which was zero except at a single point (i.e., it's a Kronecker delta: $H_i = \delta_{i,j} = 1$ if $i = j$ and zero otherwise; here a different j is chosen for each plot). Starting from the top and working down, we choose j to be the following samples: the first, the second, Nyquist-1, Nyquist, Nyquist+1, the last. We can see that the first sample in frequency domain is associated with the zero-frequency or DC component of a signal and that the frequency increases until we reach Nyquist, which is in the middle of the array.

Next, in Figure 2.5, we show at the top an input time series consisting of a pure sinusoid (left) and the real part of its DFT. Next we add some random noise to this signal. On the left in the middle plot is the real part of the noisy signals DFT. Finally, at the bottom, we show a Gaussian which we convolve with the noisy signal in order to attenuate the frequency components in the signal. The real part of the inverse DFT of this convolved signal is shown in the lower right plot.

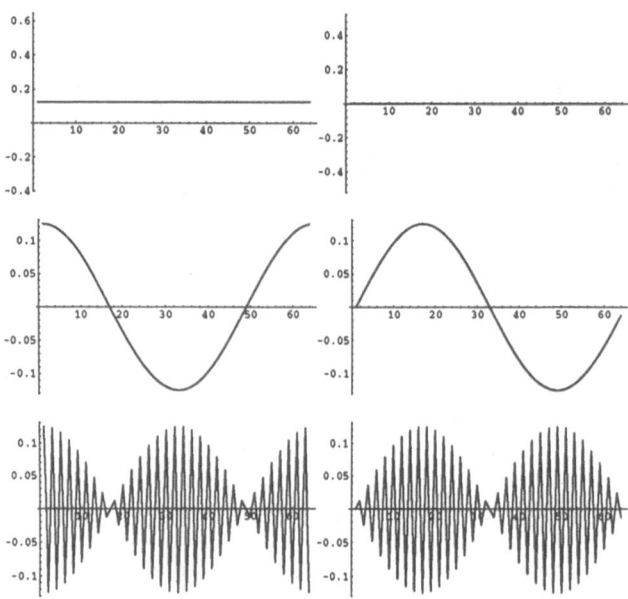

Fig. 2.3. The real (left) and imaginary (right) parts of three length 64 time series, each associated with a Kronecker delta frequency spectrum. These time series are reconstructed from the spectra by inverse DFT. At the top the input spectrum is $\delta_{i,0}$, in the middle $\delta_{i,1}$, and at the bottom, $\delta_{i,64/2-1}$.

2.4.2 Trigonometric Interpolation

There is yet is another interpretation of the DFT that turns out to be quite useful in approximation theory. Imagine that we have N samples spaced evenly along the x axis:

$$(x_k, f_k) \qquad k = 0, 1, \ldots, N-1$$

where the f_k may be complex, and $x_k = \frac{2\pi k}{N}$. Then there exists a unique polynomial

$$p(x) = \beta_0 + \beta_1 e^{ix} + \beta_2 e^{i2x} + \cdots + \beta_{N-1} e^{i(N-1)x}$$

such that $p(x_k) = f_k$. Moreover, the β_j are given by

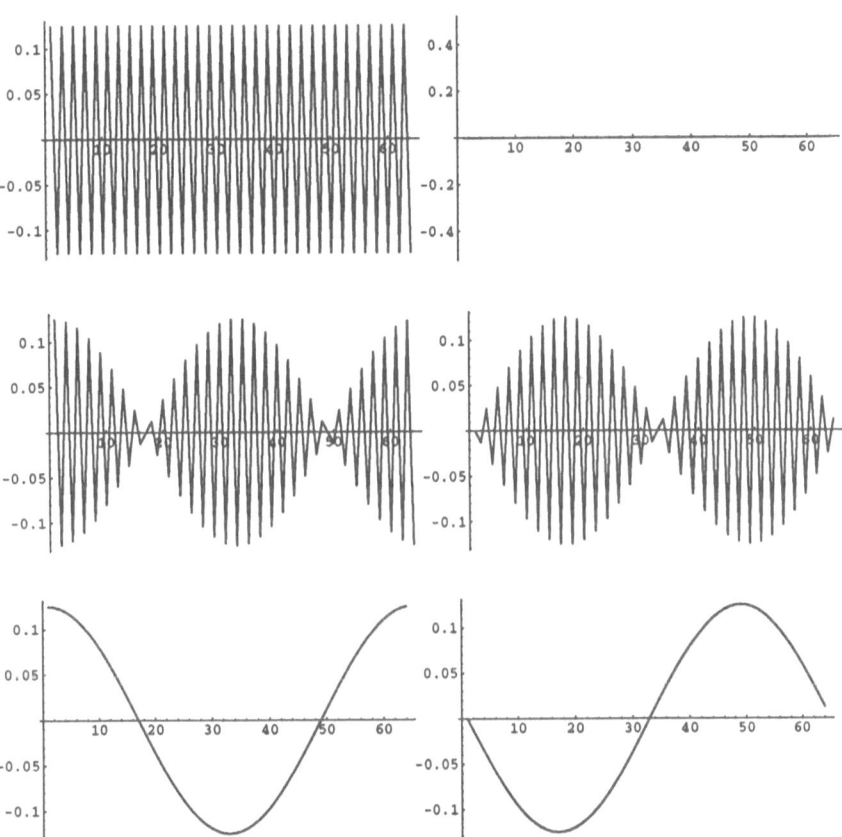

Fig. 2.4. The real (left) and imaginary (right) parts of three time series of length 64, each associated with a Kronecker delta frequency spectrum. These time series are reconstructed from the spectra by inverse DFT. At the top the input spectrum is $\delta_{i,64/2}$, in the middle $\delta_{i,64/2+1}$, and at the bottom $\delta_{i,64}$.

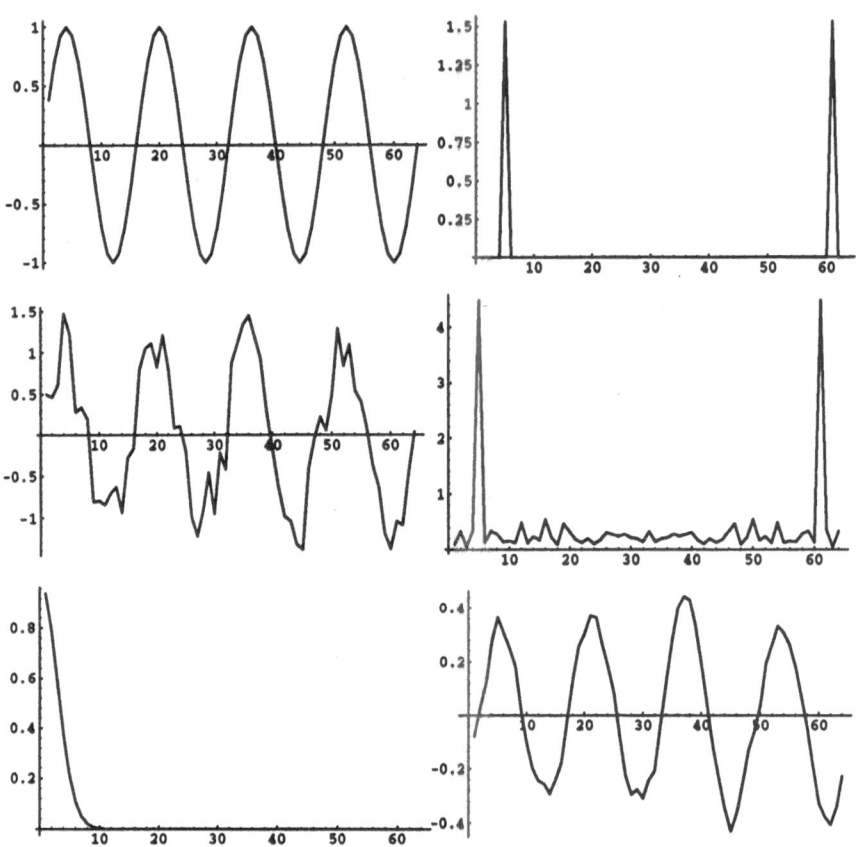

Fig. 2.5. The top left plot shows an input time series consisting of a single sinusoid. In the top right we see the real part of its DFT. Note well the wrap-around at negative frequencies. In the middle we show the same input sinusoid contaminated with some uniformly distributed pseudorandom noise and its DFT. At the bottom left, we show a Gaussian time series that we will use to smooth the noisy time series by convolving it with the DFT of the noisy signal. When we inverse DFT to get back into the "time" domain we get the smoothed signal shown in the lower right.

$$\beta_j = \frac{1}{N} \sum_{k=0}^{N-1} f_k e^{-2\pi ijk/N}. \qquad (2.41)$$

Thus by performing a DFT on N data we get a trigonometric interpolating polynomial of order $N - 1$.

Exercises

2.1 Verify Equation (2.11).

2.2 Find the Fourier transform pairs of the following, for a generic function $g(t)$ and where a, b, t_0, ω_0 are arbitrary scalars. $g(at)$. $1/|b| g(t/b)$. $g(t - t_0)$. $g(t)e^{-i\omega_0 t}$.

2.3 Define the convolution of two functions g and h as

$$(g * h)(t) \equiv \int_{-\infty}^{\infty} g(\tau)h(t - \tau) \, d\tau$$

Show that the convolution of two functions is equal to the product of their Fourier transforms.

2.4 The correlation of two functions is just like the convolution except that the minus sign becomes a plus. Show that the correlation of two functions is equal to the product of the Fourier transform of the first times the complex conjugate of the Fourier transform of the second.

2.5 Show that the autocorrelation of a function (its correlation with itself) is equal to the absolute value squared of the Fourier transform of the function (Wiener-Khinchin Theorem).

2.6 Show that if a function is imaginary and even, its Fourier transform is also imaginary and even. (A function is said to be even if $f(-x) = f(x)$ and odd if $f(-x) = -f(x)$.

2.7 What is the Fourier transform of $\frac{\mu}{\sqrt{\pi}} e^{-\mu^2 x^2}$?

2.8 What is the Nyquist frequency for data sampled at 4 ms (i.e., .004 seconds/sample).

2.5 Computer Exercise: II

In this exercise we're going to get a little experience playing with real data. You will be given instructions on how to ftp a shot record from the collection described in Oz Yilmaz's book *Seismic Data Processing*.[4] This book has data from all over the world, recorded with many different types of sources, both marine and land.

The subject of this exercise is shot record number 25. Here is its header file:

```
1 inner offset zero trace (49) added; source at trace 49
phone spacing 50 meters, inner offset 50 meters
survey=land area=Alberta source=Dynamite spread=split
n1=2000 d1=.002 f1=.002 label1=''Time (sec)''
n2=97 d2=.050 f2=-2.400 label2=''Offset (km)''
n3=1
title=''Oz 25''
plottpow=3.191650
```

The SU plotting programs will accept files of this form as command line arguments. Thus once you get your segy version of the data (see below) (call it, for example, data.segy) you could do the following:

```
suximage < data.segy par=oz25.H
```

or

```
supsimage < data.segy par=oz25.H > data.segy.ps
```

Exercises

2.9 The first thing you need to do is put the data in segy format so that you can use all the SU code on it. This is easy, just use suaddhead. You only have to tell suaddhead to put in the number of samples (ns=something) and (if the data are other than 4 mil) the sample interval. Look closely at the example in the selfdoc of suaddhead, however, to see the way in which sample intervals are specified.

2.10 Now that you've got an segy data set you can look at it with various programs such as supswigb, supswigp, supsimage (and displayed using ghostscript in X windows or open in NeXTStep) or suximage in X windows. However, you'll

---

[4] At present these are available via ftp from hilbert.mines.colorado.edu (138.67.12.63) in the directory pub/data.

immediately notice that the traces are dominated by ground roll (Rayleigh waves) near the origin. In fact, that's about all you'll see. So that you can see the other events too, try using the "perc" option in the plotting. For example:

```
supswigb < data perc=90
```

where "data" is the name of your segy data set.

2.11 Ground roll is a relatively low-frequency kind of "noise" so try using sufilter on your data to get rid of it. Play around with the frequencies and see what pass band works best for eliminating these events.

2.12 OK, now that you've eliminated the ground roll, let's try eliminating the first arrivals. These are the diagonal (linear moveout) events associated with head waves propagating along the base of the near-surface weathering layer(s). Here we can just use a plain mute to kill these events. The program sumute will do this for you. It expects an array of x locations and t values. For example if we wanted to mute straight across a section having 200 traces, at a time of 2 seconds, we might use something like

```
sumute < data xmute=1,200 tmute=2.0,2.0 key=tracl > out
```

The use of the keyword *tracl* lets us input x locations in terms of trace numbers rather than offsets. If you'd rather do things in terms of offset, however, that's up to you.

2.13 Finally boost the amplitude of the deeper reflections by running sugain over the data. In the figure below, I've used the tpow option. You might have success with other options as well.

Look around at the other SU programs and see if there are any things you'd like to do to help clean up the data. What you want to end up with is the section with the best-looking reflections (i.e., hyperbolic events). Be sure you keep track in which order you do things: seismic processing operators are definitely NOT commutative.

If you're using X windows, suximage is nice. You can enlarge an image by pushing the left mouse button and dragging a box over the zone you want to enlarge. To get back to unit magnification, click the left mouse button. You can view postscript images in X with gs, which stands for ghostscript.[5] So typing gs file.ps will display it on your screen. You can send postscript files to be printed on the laser printer just by typing lpr file.ps, as you would for an ordinary file, provided your default

---

[5] Ghostscript is freely available over Internet from various archive locations, if it's not already on your system. Ghostview is a free X11 user interface for ghostscript. You can view multi-page documents with it.

printer is equiped for postscript. In Figure 2.6 you will see my attempt at cleaning up these data. The figure illustrates simple straight-line mutes. Perhaps you can do better. These plots suffer somewhat since I've used supswigb instead of supswigp. The latter does fancy polygon fill drawing, but at the expense of a huge postscript file. That's OK for you since once you print if off you can delete the postscript. But for these notes we must keep a copy of all the postscript plots around, so we've chosen to use the more economical supswigb.

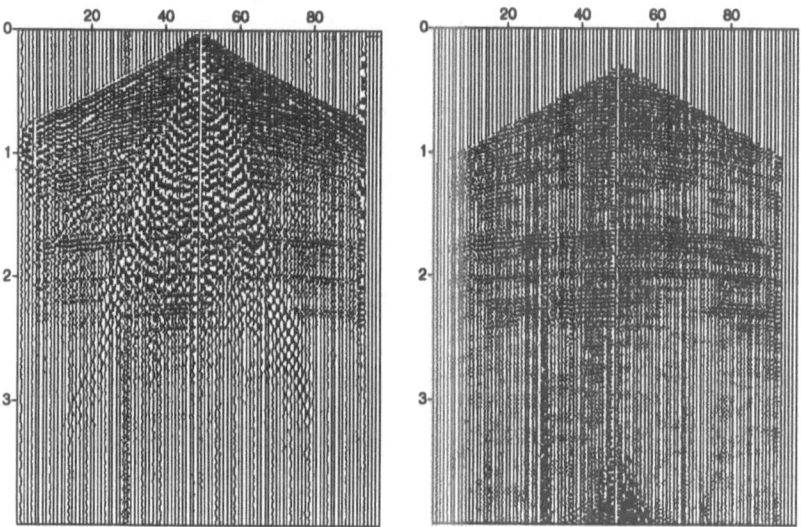

Fig. 2.6. Shot record 25 and a cleaned up version

Sinc Function Interpolation
via
Mathematica

**The sampling theorem says that if we multiply
the Nyquist-sampled values of a function by a
suitably scaled sinc function, we get the whole
(band-limited) function**

■ **Here is the first example. 25 Hz data on the time
interval [0,.5]. First, let's look at two different sinc
functions.**

```
fc = 25.;
sinc[t_,n_] = Sin[Pi ( 2. fc t - n)]/
    (Pi ( 2. fc t -n));
plot1 = Plot[sinc[t,10],{t,0.,.5},PlotRange->All];
```

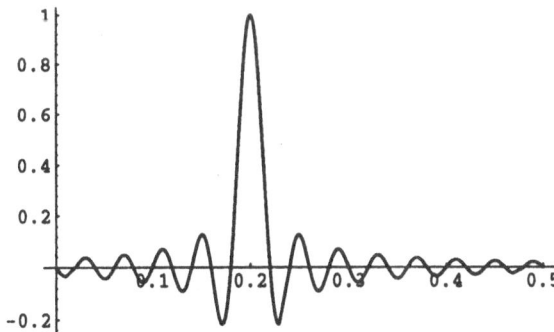

```
plot2 = Plot[sinc[t,11],{t,0.,.5},PlotRange->All];
```

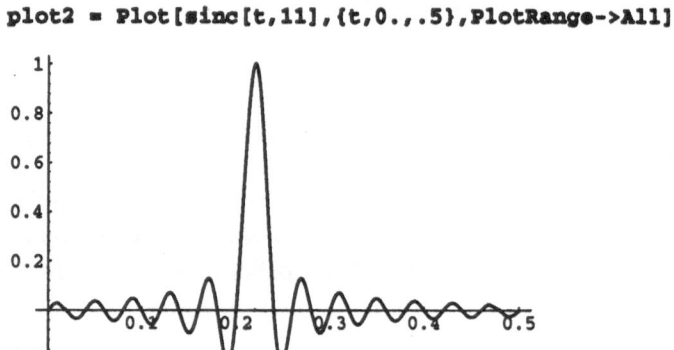

■ **The key point is that when we multiply each of these by the function values and add them up, the wiggles cancel out, leaving an interpolating approximation to the function.**

```
Show[plot1,plot2]
```

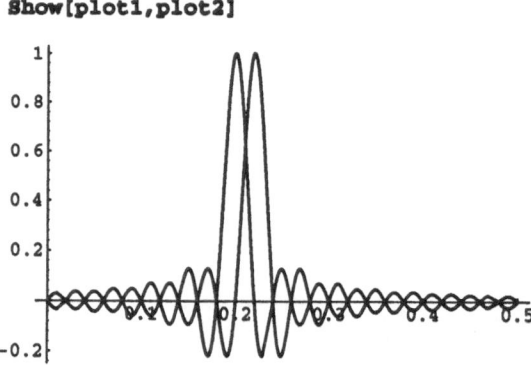

■ Suppose we are trying to approximate the constant function f(x) = 1. So we multiply the sinc functions by 1.

```
myapprox[t_] = sinc[t,11] + sinc[t,10]
```

$$\frac{Sin[Pi\ (-11\ +\ 50.\ t)]}{Pi\ (-11\ +\ 50.\ t)}\ +\ \frac{Sin[Pi\ (-10\ +\ 50.\ t)]}{Pi\ (-10\ +\ 50.\ t)}$$

```
Plot[myapprox[t],{t,0.,.5},PlotRange->All]
```

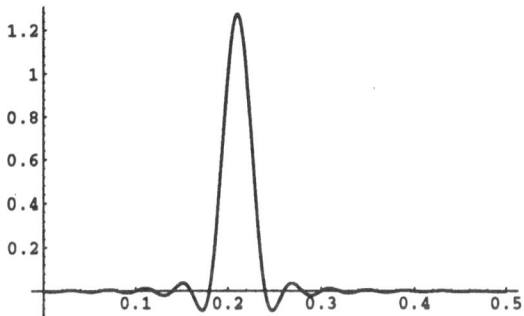

■ Extend this. Sum a lot of these up.

```
myapprox[t_] = Sum[sinc[t,n],{n,1,25}];
Plot[myapprox[t],{t,.01,.49},PlotRange->{{.1,.4},{0,2}}]
```

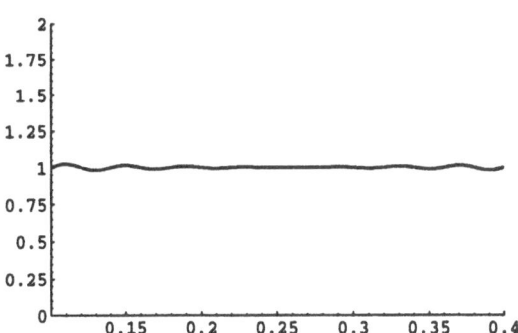

3. Equations of Motion for the Earth [1]

In the course of our study of migration methods we will encounter many different wave equations. How can that be, you say? I know the wave equation, it's

$$\nabla^2 \psi - \frac{1}{c^2} \psi = 0$$

So what are you trying to pull?

As you will see shortly, this is only an approximate wave equation. Implicit in it are assumptions about the frequency of the waves of interest, the smoothness of the material properties, and others. Further, migration involves many kinds of manipulations of the data which affect the approximate wave equations that we use. For example, since stacking attenuates multiples, if we're doing post stack migration we need a wave equation which does not give rise to spurious multiples; something which certainly cannot be said for the above equation. Therefore we thought it would be useful to give a careful derivation of the "wave equation" with all of the (well, almost all of them) assumptions up front. If derivations of the wave equation are old hat to you, feel free to skip ahead to Section 4.3 on page 61

So, let us begin by deriving the equations governing the infinitesimal, elastic-gravitational deformation of an Earth model which is spherically symmetric and non-rotating, has an isotropic and perfectly elastic incremental constitutive relation, and has an initial stress field which is perfectly hydrostatic. This is kind of a lot at one time, and some of these complications are not important in many seismological applications. It is better in the long run, however, to do the most complicated case initially and then simplify it later.

We characterize the Earth's equilibrium (non-moving) state by its equilibrium stress field, $\mathbf{T}_0$, its equilibrium density field, ρ_0, and its equilibrium gravitational potential field, ϕ_0. These are related by

$$\nabla \cdot \mathbf{T}_0(\mathbf{x}) = \rho_0(\mathbf{x}) \nabla \phi_0(\mathbf{x}) \tag{3.1}$$

---

[1] This chapter and the next two, on elastic wave equations and rays, were co-written with Martin Smith of New England Research.

and by Poisson's equation

$$\nabla^2 \phi_0(\mathbf{x}) = 4\pi G \rho_0(\mathbf{x}) \tag{3.2}$$

plus some boundary conditions. We also suppose that $\mathbf{T}_0$ is purely hydrostatic. Then

$$\mathbf{T}_0 = -p_0 \mathbf{I} \quad p_0 = \text{equilibrium pressure field.} \tag{3.3}$$

This is reasonable if the Earth has no significant strength. The equilibrium equations are then

$$\nabla p_0 = -\rho_0 \nabla \phi_0 \tag{3.4}$$
$$\nabla^2 \phi_0 = 4\pi G \rho_0 \tag{3.5}$$

Suppose, now, that the Earth is disturbed slightly from its equilibrium configuration. We have to be a little careful with our coordinates. We express particle motion in the usual form

$$\mathbf{r}(\mathbf{x}, t) = \mathbf{x} + \mathbf{s}(\mathbf{x}, t) \tag{3.6}$$

where $\mathbf{x}$ can be regarded as the particle's equilibrium position.

We express the time-dependent stress, density, and gravitational potential at a fixed spatial point $\mathbf{r}$ as

$$\phi(\mathbf{r}, t) = \phi_0(\mathbf{r}) + \phi_1(\mathbf{r}, t) \tag{3.7}$$
$$\rho(\mathbf{r}, t) = \rho_0(\mathbf{r}) + \rho_1(\mathbf{r}, t) \tag{3.8}$$
$$\mathbf{T}(\mathbf{r}, t) = \mathbf{T}_0(\mathbf{r}) + \mathbf{T}_1(\mathbf{r}, t) \tag{3.9}$$
$$= -p_0(\mathbf{r})\mathbf{I} + \mathbf{T}_1(\mathbf{r}, t) \tag{3.10}$$

These equations define the "incremental" quantities ϕ_1, ρ_1, and $\mathbf{T}_1$. Note that the incremental quantities are defined as differences from the equilibrium values at the observing point; they are not, in general, the changes experienced by a material particle. We also need the particle acceleration

$$d_t \mathbf{v}(\mathbf{r}, t) = \partial_t^2 \mathbf{s}(\mathbf{x}, t) \text{ to first order in } \mathbf{s} \tag{3.11}$$

The equations of motion are, then to first order

$$\rho_0 \partial_t^2 \mathbf{s} = -\rho_0 \nabla \phi_0 = \rho_0 \nabla \phi_1 - \rho_1 \nabla \phi_0 - \nabla p_0 + \nabla \cdot \mathbf{T}_1 \tag{3.12}$$

using the fact that the incremental quantities are all of order $\mathbf{s}$ and keeping only terms through the first order. If we subtract the equilibrium condition we are left with

$$\rho_0 \partial_t^2 \mathbf{s} = -\rho_0 \nabla \phi_1 - \rho_1 \nabla \phi_0 + \nabla \cdot \mathbf{T}_1 \tag{3.13}$$

Each term in this equation is evaluated at the spatial point $\mathbf{r}$, but the equation governs the acceleration of the material packet which may be at $\mathbf{r}$ only for the single instant of time, t.

We also must have

$$\partial_t(\rho_0 + \rho_1) + (\rho_0 + \rho_1)\nabla \cdot (\partial_t \mathbf{s}) + (\partial_t \mathbf{s}) \cdot \nabla(\rho_0 + \rho_1) = 0 \tag{3.14}$$

so to first order

$$\partial_t \rho_1 + \rho_0 \nabla \cdot (\partial_t \mathbf{s}) + (\partial_t \mathbf{s}) \cdot \nabla \rho_0 \quad = \quad 0 \tag{3.15}$$
$$\partial_t \{\rho_1 + \nabla \cdot (\rho_0 \mathbf{s})\} \quad = \quad 0 \tag{3.16}$$
$$\rho_1 \quad = \quad -\nabla \cdot (\rho_0 \mathbf{s}) \tag{3.17}$$
$$\text{since } \rho_1 = 0 \quad \text{if} \quad \mathbf{s} = 0. \tag{3.18}$$

Further,

$$\nabla^2(\phi_0 + \phi_1) \quad = \quad 4\pi G(\rho_0 + \rho_1) \tag{3.19}$$
$$\nabla^2 \phi_1 = 4\pi G \rho_1 \quad = \quad -4\pi G \nabla \cdot (\rho_0 \mathbf{s}) \tag{3.20}$$

The last bit we have to sort out is how to compute $\mathbf{T}_1$, the incremental stress at the point $\mathbf{r}$. This is more complicated than it might appear because $\mathbf{T}_1$ is *not* simply the product of the local infinitesimal strain tensor and the local elastic tensor. The reason we have this problem is the presence of an initial stress field, $\mathbf{T}_0$ which is not infinitesimal but is of order zero in the displacement.

Suppose, for example, we keep our attention on the point $\mathbf{r}$ while we rigidly translate the Earth a small amount. After the translation, we will be looking at the same spatial point $\mathbf{r}$ but at a different piece of matter. This new piece will in general have a different initial stress field than our old piece. If we translated the Earth an amount $\mathbf{d}$ we would perceive an incremental stress of

$$\mathbf{T}_1(\mathbf{r}, t) = \mathbf{T}_0(\mathbf{r} - \mathbf{d}) - \mathbf{T}_0(\mathbf{r}) \neq 0 \text{ in general.} \tag{3.21}$$

However, a rigid translation produces no strain, as we well know, and thus cannot cause any elastic stress. Thus $\mathbf{T}_1$, above, has nothing whatever to do with elasticity or strain but is solely a result of the initial stress field.

To see how to handle this problem, consider the stress tensor at a material particle, which we express as

$$-p_0(\mathbf{x})\mathbf{I} + \mathbf{T}_m \text{ at } \mathbf{r} = \mathbf{x} + \mathbf{s}(\mathbf{x}, t) \tag{3.22}$$

We have expressed the total stress at the particle $\mathbf{x}$, which is now located at $\mathbf{r}$, as the initial stress *at the particle* plus an increment $\mathbf{T}_m$. (Note that we can use either $\mathbf{T}_m(\mathbf{x})$ or $\mathbf{T}_m(\mathbf{r})$ since $\mathbf{T}_m$, as we will see, is first-order in $\mathbf{s}$. We cannot use $p_0(\mathbf{r})$, however, if we want $\mathbf{T}_m$ to mean what we claim it to.) We now claim

$$\mathbf{T}_m = \lambda(\nabla \cdot \mathbf{s})\mathbf{I} + \mu(\nabla\mathbf{s} + \mathbf{s}\nabla) \tag{3.23}$$

because the only way to alter the stress at a particle is by straining the continuum. Remember that the stress tensor describes the forces between parts of the continuum across intervening surfaces. These forces, in turn, are exactly a result of the material being distorted out of its preferred state. The expression for $\mathbf{T}_m$, in fact, defines what we mean by an "incremental constitutive relation."

If $-p_0(\mathbf{x})\mathbf{I} + \mathbf{T}_m$ is the stress tensor at a particle, then the stress tensor at a point in space is simply

$$(\mathbf{I} - \mathbf{s} \cdot \nabla)(-p_0\mathbf{I} + \mathbf{T}_m) = \mathbf{T}_0 + \mathbf{T}_1 \tag{3.24}$$

which is just a Taylor series. To first order, we must have

$$\mathbf{T}_1 = \mathbf{T}_m + (\mathbf{s} \cdot \nabla p_0)\mathbf{I} \tag{3.25}$$

So the full set of linearized equations is

$$\rho_0 \partial_t^2 \mathbf{s} = -\rho_0 \nabla\phi_1 - \rho_1 \nabla\phi_0 + \nabla \cdot \mathbf{T}_m + \nabla \cdot (\mathbf{s} \cdot \nabla p_0) \tag{3.26}$$
$$\rho_1 = -\nabla \cdot (\rho_0 \mathbf{s}) \tag{3.27}$$
$$\nabla^2\phi_1 = 4\pi G \rho_1 \tag{3.28}$$
$$\mathbf{T}_m = \lambda(\nabla \cdot \mathbf{s})\mathbf{I} + \mu(\nabla\mathbf{s} + \mathbf{s}\nabla) \tag{3.29}$$
$$\nabla p_0 = -\rho_0 \nabla\phi_0 \tag{3.30}$$

Remember that we have assumed that the Earth

- was not rotating

- had a hydrostatic prestress field

- was self-gravitating

- was spherically symmetric

- had an isotropic, perfectly elastic incremental constitutive relation

Remember, too, that in formulating these equations we encountered a situation in which the distinction between spatial and material concepts was crucial. It is often said that in linear (infinitesimal) elasticity there is no difference between

Eulerian (spatial) and Lagrangian (material) descriptions. That is wrong. There is no difference in problems which do not have zeroeth order initial stress (or other) fields.

3.1 Computer Exercise: III

The purpose of this exercise is to make sure you are thoroughly familiar with segy headers and to prepare you for a later exercise in which you will need to process a blind data set. You will begin by making some synthetic data with susynlv. Then you will add whatever header information is necessary for you to complete the following processing flow:

– Do stacking velocity analysis using velan.

– NMO correct and stack the data to produce a zero-offset section.

– Migrate the data with sugazmig.

– Repeat the migration and velocity analysis until you are satisfied with the image.

These are just the main steps. There will be other small steps such as figuring out which headers are required, sorting the data and so on.

For creating the synthetic data with susynlv use the following description of the reflectors:

```
ref=''0,.25;10.0,.25''  \
ref=''0,.35;5.0,.5;10.0,.35''  \
ref=''0,.75;10.,1.0''  \
ref=''0,1.5;10.,1.5''
```

The sample interval is 4 ms, the shot spacing is 50, the receiver spacing is 50. You can do either a split spread or single ended survey. For velocities, use v00=1.0 dvdz=2,0. dvdx=0. Record enough samples so that you can see the deepest reflector.

A quick way of seeing what's going on might be to pick the near offset traces off of each shot record and take these as a zero-offset section.

Susynlv will dump all the traces, shot by shot into a single file. You can examine all of traces simultaneously with suximage, or you can use suwind to pick off a smaller number to look at. Or you can use the zoom cababilities of suximage to zoom in on individual shot records.

3.2 Computer Exercise: III — An Example Solution

Here is one solution to Exercise III contributed by Tagir Galikeev and Gabriel
Alvarez. At the time this was done, we had not begun using SUB in the class. Try
writing a SUB script to do the pre-processing.

```
#--- Create Data ------------------------------
susynlv > data.su nt=251 dt=0.008 nxs=200 dxs=0.05 \
     nxo=60 dxo=0.05 fxo=-1.475 ref="0,0.25;10,0.25" \
     ref="0,0.35;5,0.5;10,0.35" ref="0,0.75;10,1.0"  \
     ref="0,1.5;10,1.5" v00=1.0 dvdz=2.0 verbose=1   \

#--- Reformat to SEGY --------------------------
segyhdrs <data.su
segywrite <data.su tape=data.segy

#--- Edit ( Trace Kills on the Edges of the Model )
#--- use two temporary files
sukill <data.su min=1 count=30 |
sukill min=61 count=29 |
sukill min=121 count=28 |
sukill min=181 count=27 |
sukill min=241 count=26 |
sukill min=301 count=25 |
sukill min=361 count=24 |
sukill min=421 count=23 |
sukill min=481 count=22 |
sukill min=541 count=21 |
sukill min=601 count=20 |
sukill min=661 count=19 |
sukill min=721 count=18 |
sukill min=781 count=17 |
sukill min=841 count=16 |
sukill >temp1 min=901 count=15
sukill <temp1 min=961 count=14 |
sukill min=1021 count=13 |
sukill min=1081 count=12 |
sukill min=1141 count=11 |
sukill min=1201 count=10 |
sukill min=1261 count= 9 |
sukill min=1321 count= 8 |
sukill min=1381 count= 7 |
sukill min=1441 count= 6 |
sukill min=1501 count= 5 |
```

```
sukill min=1561 count= 4 |
sukill min=1621 count= 3 |
sukill min=1681 count= 2 |
sukill >temp2 min=1741 count= 1
sukill <temp2 min=10260 count= 1 |
sukill min=10319 count= 2 |
sukill min=10378 count= 3 |
sukill min=10437 count= 4 |
sukill min=10496 count= 5 |
sukill min=10555 count= 6 |
sukill min=10614 count= 7 |
sukill min=10673 count= 8 |
sukill min=10732 count= 9 |
sukill min=10791 count=10 |
sukill min=10850 count=11 |
sukill min=10909 count=12 |
sukill min=10968 count=13 |
sukill min=11027 count=14 |
sukill >temp1 min=11086 count=15
sukill <temp1 min=11145 count=16 |
sukill min=11204 count=17 |
sukill min=11263 count=18 |
sukill min=11322 count=19 |
sukill min=11381 count=20 |
sukill min=11440 count=21 |
sukill min=11499 count=22 |
sukill min=11558 count=23 |
sukill min=11617 count=24 |
sukill min=11676 count=25 |
sukill min=11735 count=26 |
sukill min=11794 count=27 |
sukill min=11853 count=28 |
sukill min=11912 count=29 |
sukill >data.edit.su min=11971 count=30

#--- Geometry Assignment & CDP Sort ------------------
sushw <data.edit.su key=sx a=130.5 b=0 c=1.0 d=0 j=60 |
sushw key=offset a=-1475 b=50 j=60 |
sushw key=gx a=101 b=1 c=1 d=0 j=60 |
suchw key1=cdp key2=gx key3=sx a=-1 b=1 c=1 |
susort >data.geom.cdp.su cdp offset

#--- Perform Velocity Analysis ----------------------
#--- Analysis on CDP 290
```

```
suwind <data.geom.cdp.su key=cdp min=290 max=290 |
suvelan >semblance290.su fv=500.0 dv=100.0
#--- Analysis on CDP 390
suwind <data.geom.cdp.su key=cdp min=390 max=390 |
suvelan >semblance390.su fv=500.0 dv=100.0
#--- Analysis on CDP 490
suwind <data.geom.cdp.su key=cdp min=490 max=490 |
suvelan >semblance490.su fv=500.0 dv=100.0
#--- Analysis on CDP 590
suwind <data.geom.cdp.su key=cdp min=590 max=590 |
suvelan >semblance590.su fv=500.0 dv=100.0


#--- Apply NMO Correction, Stack  --------------------
sunmo <data.geom.cdp.su cdp=290,390,490,590   \
      vnmo=1250,1500,1800,2400                \
      tnmo=0.47,0.70,1.0,1.45                 \
      vnmo=1250,1400,1800,2350,3000           \
      tnmo=0.47,0.57,1.0,1.4,1.57             \
      vnmo=1250,1450,1850,2400                \
      tnmo=0.47,0.68,1.0,1.42                 \
      vnmo=1250,1400,1900,2350                \
      tnmo=0.47,0.6,1.07,1.4 |
sustack >stack.su


#--- Gazdag Migration ---------------------------------
sugazmig <stack.su tmig=0.0,2.0 vmig=1000.,5000. dx=25 |
supsimage >stack.mig.ps \
      title="Migrated Stack with True Velocities"
```

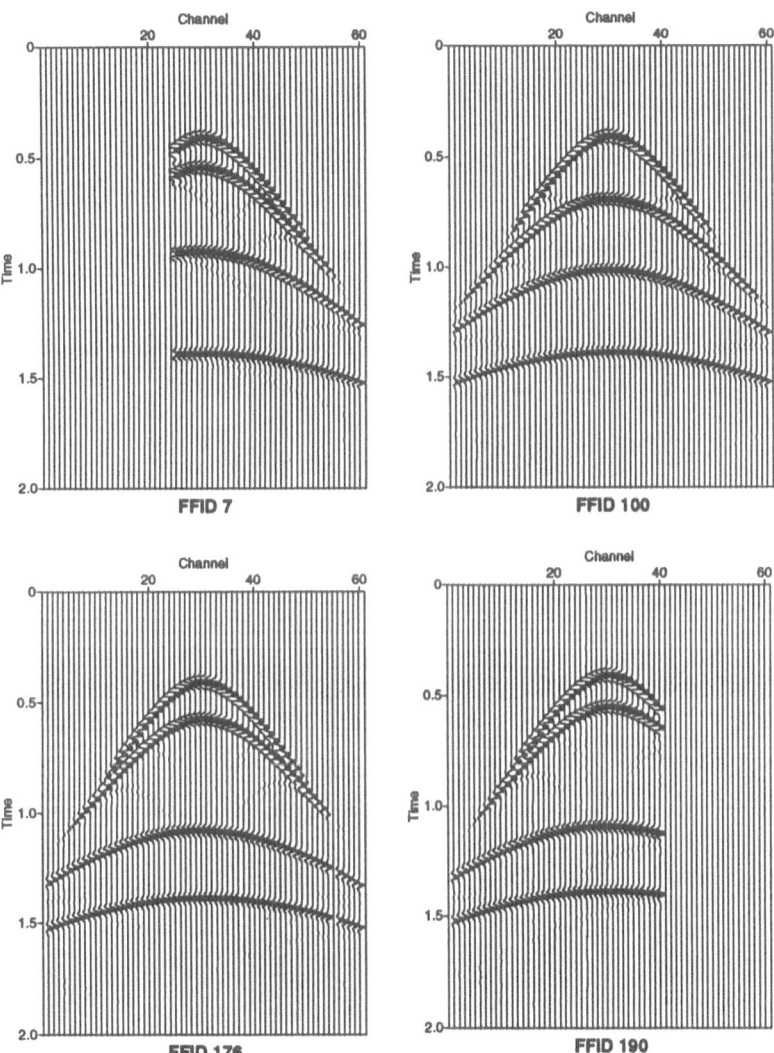

Fig. 3.1. The raw data

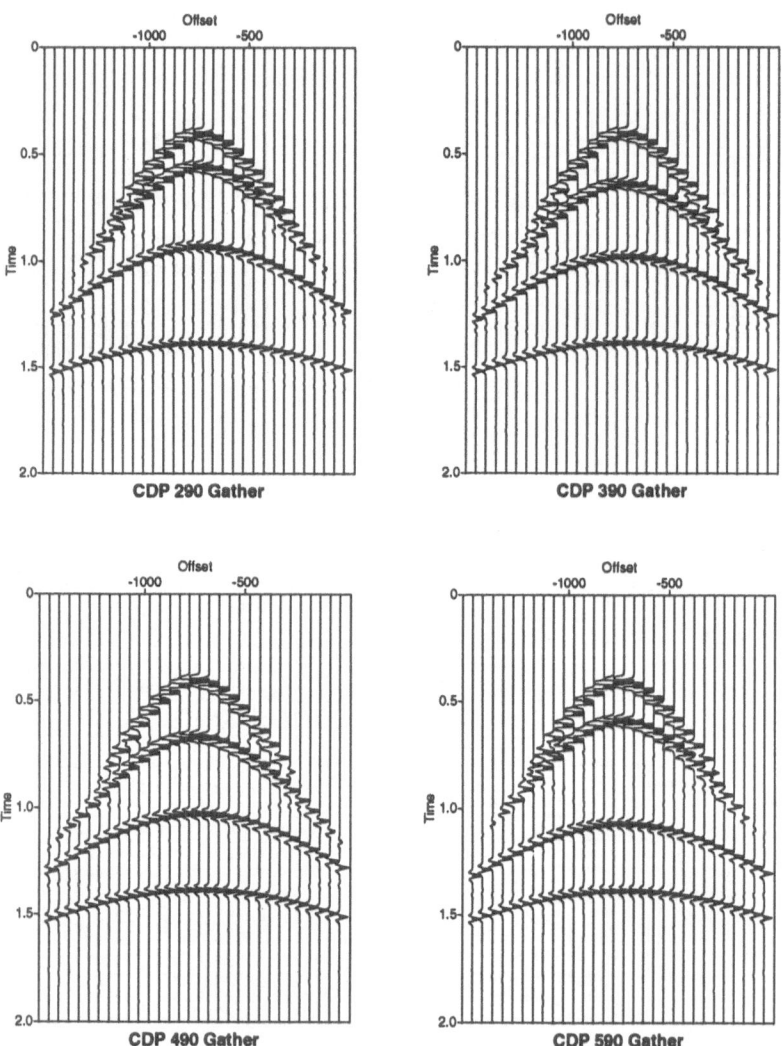

Fig. 3.2. CDP sorted data

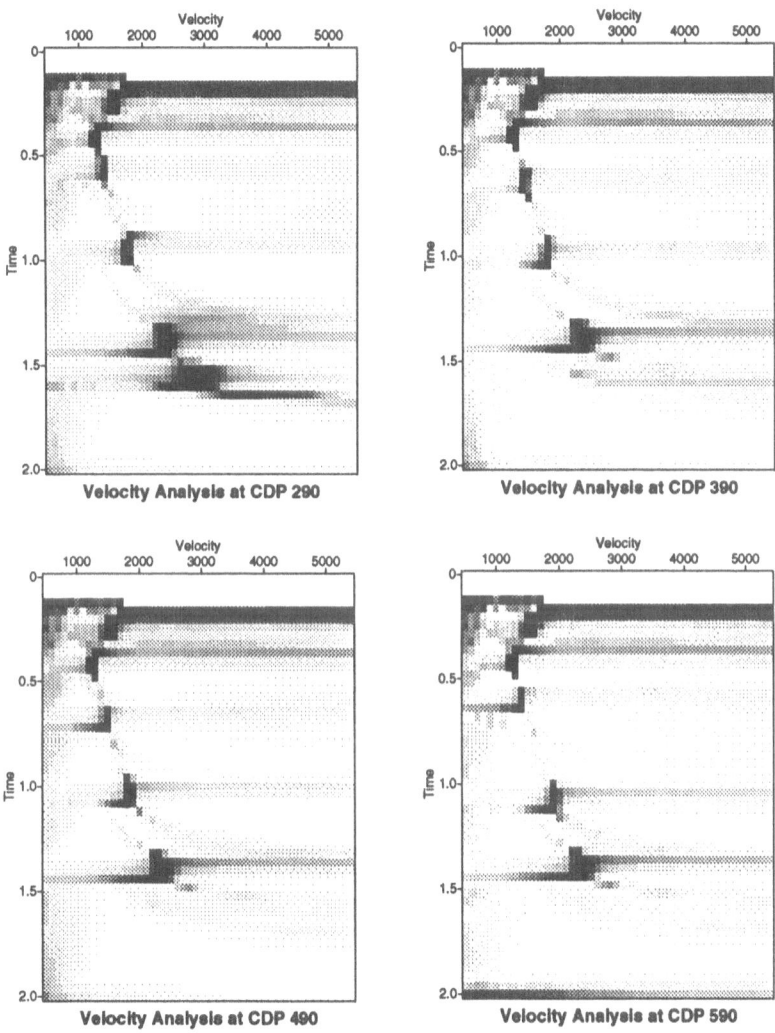

Fig. 3.3. Velocity analysis panels

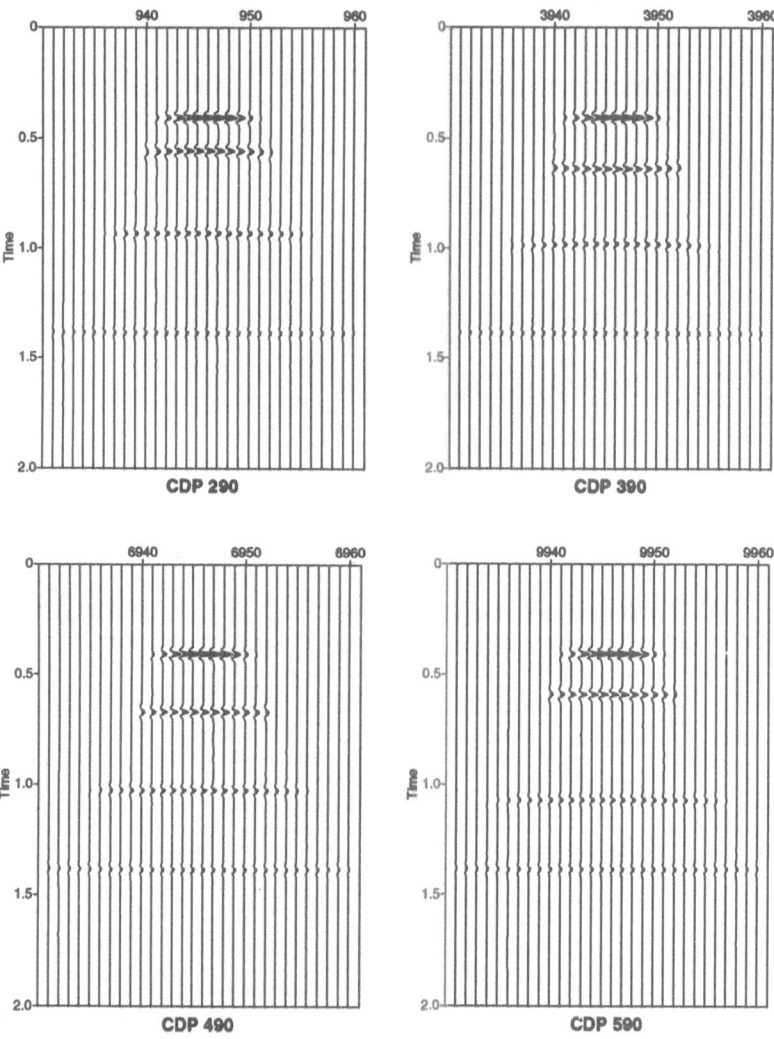

Fig. 3.4. Moveout corrected traces

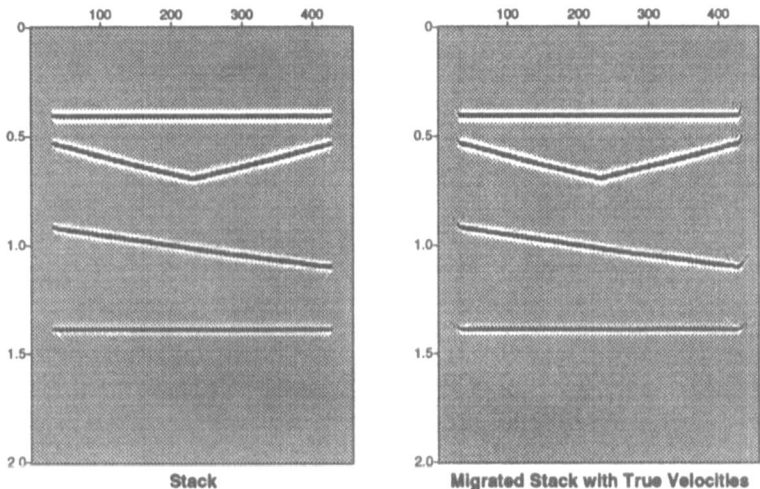

Fig. 3.5. The final stack and zero-offset migration

4. Elastic Wave Equations

4.1 Gravity Versus Elasticity

Our most recent version of the equations of motion,

$$\rho_0 \partial_t^2 \mathbf{s} = -\rho_1 \nabla \phi_0 - \rho_0 \nabla \phi_1 + \nabla(\mathbf{s} \cdot p_0) + \nabla \cdot \mathbf{T}_m \qquad (4.1)$$
$$+ \quad \text{four more relations}$$

is a little overwhelming and we might hope that not every geophysical application requires solving the full-blown system.

There is a very important class of problems requiring the simultaneous solution of the entire system, namely the elastic-gravitational normal mode problem. On the other hand, as we will crudely demonstrate here, at the higher range of frequencies of seismological interest, a large portion of the equation of motion can be neglected.

To see this, we shall have to make premature use of the notion of a propagating seismic wave. Suppose that some sort of wave with angular frequency ω and wave-vector $\mathbf{k}$ is propagating through the Earth. Then, being deliberately vague,

$$\mathbf{s} \text{ goes like } e^{i(\mathbf{k}\cdot\mathbf{r}-\omega t)} \text{ more or less.} \qquad (4.2)$$

Let us consider how the various terms in the momentum equation scale with ω and $k = |\mathbf{k}|$. We neglect the spatial variation of ρ, λ, and μ and use S to denote the scale size of displacement:

$$\rho_0 \partial_t^2 \mathbf{s} \approx \omega^2 \rho_0 S \qquad (4.3)$$
$$\rho_1 \approx \rho_0 k S \qquad (4.4)$$
$$\rho_1 \nabla \phi_0 \approx \rho_0 k g S \text{ where } g = \text{gravity} \qquad (4.5)$$
$$\nabla(\mathbf{s} \cdot \nabla p_0) \approx \rho_0 k g S \qquad (4.6)$$
$$\nabla \cdot \mathbf{T} \approx k^2 \lambda \text{ or } k^2 \mu \qquad (4.7)$$

(We have skipped over ϕ_1, for simplicity. Generally speaking, ϕ_1 is less significant than ϕ_0.)

The right-hand side of the momentum equation is the sum of various elastic *or* gravitational terms. Elastic terms scale like $k^2\lambda$; gravitational terms scale like $\rho_0 kg$. Thus the ratio of gravitation to elastic forces goes like

$$\frac{\text{gravity}}{\text{elasticity}} \propto \frac{\rho kg}{k^2\lambda} = \frac{\rho g}{k\lambda} \tag{4.8}$$

$$\frac{\rho}{\lambda} \approx c^2 \text{ the elastic velocity} \tag{4.9}$$

$$\frac{\rho g}{k\lambda} \approx \frac{g}{c^2 k}. \tag{4.10}$$

(I confess to reaching ahead somewhat.) We see that $k \to \infty$ results in an elastically-dominated system while $k \to 0$ results in a gravitationally-dominated system. The two influences are crudely equal when

$$\frac{g}{c^2 k} \approx 1 \tag{4.11}$$

or

$$k \approx \frac{g}{c^2} \approx \frac{10^3}{10^{12}} = 10^{-9} \text{ in the Earth.} \tag{4.12}$$

Let us further suppose that we are essentially in the elastic domain. Then $k \approx \omega/c$. This is obviously somewhat contradictory but it does give us a useful idea of where gravity becomes important. Then we can argue:

$$\frac{\omega}{c} \approx 10^{-9} \tag{4.13}$$

$$\omega \approx 10^{-3} \tag{4.14}$$

$$t = \frac{2\pi}{\omega} \approx 6000 \text{ seconds.} \tag{4.15}$$

This asserts that at periods which are short compared to 6×10^3 seconds, the equations of motion are dominated by elasticity. In the high-frequency limit, then, we can make do with

$$\rho_0 \partial_t^2 \mathbf{s} = \nabla \cdot \mathbf{T}_m \tag{4.16}$$

$$\mathbf{T}_m = \lambda(\nabla \cdot \mathbf{s})\mathbf{I} + \mu(\nabla \mathbf{s} + \mathbf{s}\nabla) \tag{4.17}$$

These are the equations of motion used in conventional short-period seismology (10^{-2} sec to 10 sec). At longer periods the influence of gravity becomes non-negligible and we must find ways to study the complete equations of motion. (A much better way to determine the point at which gravity becomes significant is to solve both versions of the equations of motion and compare the solutions.)

4.2 The Elastic Wave Equation

We will first boil down the equations of motion (for high frequencies and therefore without gravity) into a form called "the elastic wave equation." Then we will briefly review why equations of the form are called wave equations.

For sufficiently high frequencies we have

$$\rho_0 \partial_t^2 \mathbf{s} = \nabla \cdot \mathbf{T}_m \tag{4.18}$$

and

$$\mathbf{T}_m = \lambda (\nabla \cdot \mathbf{s}) \mathbf{I} + \mu (\nabla \mathbf{s} + \mathbf{s} \nabla). \tag{4.19}$$

Then

$$\rho_0 \partial_t^2 \mathbf{s} = \nabla (\lambda (\nabla \cdot \mathbf{s})) + \nabla \mu \cdot [\mu (\nabla \mathbf{s} + \mathbf{s} \nabla)] \tag{4.20}$$

$$\rho_0 \partial_t^2 \mathbf{s} = (\nabla \cdot \mathbf{s}) \nabla \lambda + \nabla \mu \cdot (\nabla \mathbf{s} + \mathbf{s} \nabla)$$

$$+ \lambda \nabla (\nabla \cdot \mathbf{s}) + \mu \nabla \cdot (\nabla \mathbf{s} + \mathbf{s} \nabla) \tag{4.21}$$

The first two terms result from spatial gradients of the elastic constants. We are going to drop them for two reasons:

1. As we could show by scale analysis, they are relatively unimportant as long as the wavelength of the solution is short compared to the scale length of changes in the elastic properties. For any continuous distribution of properties there will be some scale frequency *above which* we may ignore these terms.

2. These terms can cause an incredible amount of grief.

After surgery we have

$$\rho_0 \partial_t^2 \mathbf{s} = \lambda \nabla (\nabla \cdot \mathbf{s}) + \mu \nabla \cdot (\nabla \mathbf{s} + \mathbf{s} \nabla) \tag{4.22}$$

where λ and μ are "sufficiently constant." Since

$$\nabla \cdot (\nabla \mathbf{s} + \mathbf{s} \nabla) = 2 \nabla (\nabla \cdot \mathbf{s}) - \nabla \times \nabla \times \mathbf{s} \tag{4.23}$$

$$\rho_0 \partial_t^2 \mathbf{s} = (\lambda + 2\mu) \nabla (\nabla \cdot \mathbf{s}) - \mu \nabla \times \nabla \times \mathbf{s} \tag{4.24}$$

This is sometimes called the "elastic wave equation," although we know it to be only a high-frequency pretender.

Now express $\mathbf{s}$ as

$$\mathbf{s} = \nabla \psi + (\mathbf{r} \chi) + \nabla \times (\nabla \times \mathbf{r} \sigma) \tag{4.25}$$

where ψ, χ, and σ are scalar fields. The above expression is a representation for
s by which we mean that we believe that for any s of interest there exist scalar
fields ψ, χ, σ fulfilling the above. Further we believe that recasting our problem
in terms of these scalars will do some good.

You can verify, if you wish, that with this representation the elastic wave equation
becomes

$$\nabla(\rho_0\partial_t^2\psi - (\lambda+2\mu)\nabla^2\psi) \quad + \quad \nabla \times \mathbf{r}(\rho_0\partial_t^2\chi - \mu\nabla^2\chi) \tag{4.26}$$
$$+ \quad \nabla \times \nabla \times \mathbf{t}(\rho_0\partial_t^2\sigma - \mu\nabla^2\sigma) = 0. \tag{4.27}$$

If

$$\rho_0\partial_t^2\psi \;=\; (\lambda+2\mu)\nabla^2\psi \tag{4.28}$$
$$\rho_0\partial_t^2\chi \;=\; \mu\nabla^2\chi \text{ and} \tag{4.29}$$
$$\rho_0\partial_t^2\sigma \;=\; \mu\nabla^2\sigma \tag{4.30}$$

then s as gotten from these scalars will be a solution of the elastic wave equation.
However, as a little thought shows, we have not shown that every s satisfying
the elastic wave equation can be found from scalar fields satisfying these three
relationships. Fortunately, however, it is so and we can replace "if" by "if and
only if."

If terms involving the gradients of λ and μ appear, all of this, alas goes down the
drain and we have a much more intractable system. Qualitatively the gradient
terms couple shear and compressional motions together. When this coupling is
weak we can regard those terms as small sources, as is done in scattering theory.
If the coupling is strong enough the notions of distinct shear and compressional
waves break down.

The three scalar equations are called "scalar wave equations." This form of equation is discussed later. Observe that three perfectly valid classes of solutions can
be gotten by setting two of the scalars to zero and requiring the third to be a
non-zero solution of its respective wave equation. Let $\{\mathbf{s}_\psi\}$, $\{\mathbf{s}_\chi\}$, $\{\mathbf{s}_\sigma\}$ denote, in
the obvious way, these three sets of solutions. Because everything is linear, any
general solution to the elastic wave equation can be expressed as a linear combi-
nation of members of these three sets. Note that every $\mathbf{s}_\psi$ satisfies $\nabla \times \mathbf{s}_\psi = 0$
and every $\mathbf{s}_\chi$, $\mathbf{s}_\sigma$ satisfies $\nabla \cdot \mathbf{s}_\chi = \nabla \cdot \mathbf{s}_\sigma = 0$. Also χ and σ satisfy exactly the
same scalar wave equation.

These solutions have certain characteristic properties which we will simply describe, and not prove, here. $\mathbf{s}_\psi$ is called a "P wave" or "compressional wave." It
is a propagating dilatation with phase velocity

$$V_P = \alpha = \sqrt{\lambda+2\mu/\rho}. \tag{4.31}$$

s_χ and s_σ are called "S waves" or "shear waves." Particle motions for these waves are at right angles to the direction of propagation. These waves have a phase velocity

$$V_S = \beta = \sqrt{\mu/\rho} < V_P \tag{4.32}$$

There are two classes of shear waves simply because there are two possible orthogonal polarizations. Shear waves have no associated dilatation ($\nabla \cdot \mathbf{s} = 0$).

4.3 Scalar Wave Equations

4.3.1 Plane Waves

We will now consider why equations of the form

$$\partial_t^2 \psi = c^2 \nabla^2 \psi \tag{4.33}$$

are called "scalar wave equations." The factor c, which we suppose to be nearly constant, is called the "wave speed."

Let $\mathbf{r}$ be a position vector to any point in space, let $\hat{n}$ be a fixed direction in space and let $f(x)$ be any twice differentiable function of one variable. If

$$\psi(\mathbf{r}, t) = f(\hat{n} \cdot \mathbf{r} - ct) \tag{4.34}$$

then

$$\nabla \psi \;=\; \hat{n} f' \tag{4.35}$$
$$\nabla \cdot (\nabla \psi) \;=\; \hat{n} \cdot \hat{n} f'' = f'' \tag{4.36}$$

and

$$\partial_t \psi \;=\; -c f' \tag{4.37}$$
$$\partial_t^2 \psi \;=\; c^2 f'' \tag{4.38}$$

The wave equation becomes

$$c^2 f'' = c^2 f'' \tag{4.39}$$

which is true. Thus any ψ of the above form satisfies the wave equation. Such a ψ is constant on the planes defined by (cf. Figure 4.1)

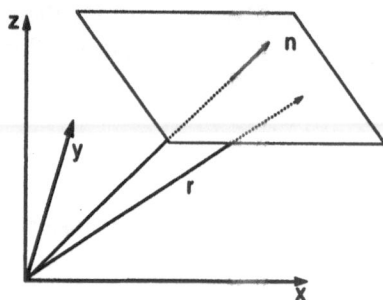

Fig. 4.1. Plane wave propagation

$$\hat{n} \cdot \mathbf{r} - ct = A, \text{ a constant} \tag{4.40}$$

Also, the plane associated with a fixed A, and therefore a particular value of ψ moves in the direction $\hat{n}$ with speed c (or to be pedantic, velocity $c\hat{n}$). Thus the pattern of ψ in space is propagating with the velocity $c\hat{n}$, which is what we expect of a "wave." Similarly $f(\hat{n} \cdot \mathbf{r} + ct)$ represents a disturbance, constant on the plane defined by $\hat{n} \cdot \mathbf{r} =$ constant, which is being propagated with velocity c in the negative $\hat{n}$ direction.

We generally think in terms of simple harmonic time dependence, and use ω to denote angular frequency. If we define the wave number

$$k = \frac{\omega}{c} = \frac{2\pi}{\lambda}, \tag{4.41}$$

which is the number of wavelengths per unit length, and the wave-number vector

$$\mathbf{k} = k\hat{n} = \frac{\omega}{c}\hat{n} \tag{4.42}$$

then

$$\hat{n} \cdot \mathbf{r} - ct = \frac{c}{\omega}(\mathbf{k} \cdot \mathbf{r} - \omega t), \tag{4.43}$$

a more conventional form.

Now if the plane wave $f(\hat{n} \cdot \mathbf{r} - ct)$ is a pressure disturbance propagating through an acoustic medium, then it is related to the particle velocity $\mathbf{v}$ via Newton's second law:

$$\rho \partial_t \mathbf{v} = -\nabla p \equiv -\nabla f(\hat{n} \cdot \mathbf{r} - ct). \tag{4.44}$$

The kinetic and potential energy densities of the plane wave are equal to $f^2/2\rho c^2$. So the total acoustic energy density is just twice this.

4.3.2 Spherical Waves

We begin with the scalar wave equation:

$$\partial_t^2 \psi = c^2 \nabla^2 \psi \tag{4.45}$$

where we assume c is constant. We have already seen that if $f(x)$ is any bounded, twice-differentiable function of a single variable, then

$$\psi(\mathbf{r}, t) = f(\hat{\mathbf{n}} \cdot \mathbf{r} - ct) \tag{4.46}$$

satisfies the wave equation for any direction $\hat{\mathbf{n}}$.

There are other simple wave solutions. In spherical coordinates, r, θ, ϕ, if we assume ψ is a function only of r and t (i.e., spherical symmetry), then

$$\nabla^2 \psi = \partial_r^2 \psi + \frac{2}{r}\partial_r \psi. \tag{4.47}$$

(Remember we are assuming that the *wave* is spherically symmetric about the origin of the coordinate system. We might imagine a purely symmetric, explosion-like source for such a wave.) The substitution

$$\psi(r, t) = \frac{1}{r}\eta(r, t) \tag{4.48}$$

produces

$$\nabla^2 \psi = \frac{1}{r}\partial_r^2 \eta \tag{4.49}$$

and the wave equation reduces to the one-dimensional wave equation

$$\partial_r^2 \eta = \frac{1}{c^2}\partial_t^2 \eta. \tag{4.50}$$

So,

$$\eta = f(r - ct) + g(r + ct) \tag{4.51}$$

and thus

$$\psi = \frac{1}{r}\{f(r - ct) + g(r + ct)\}. \tag{4.52}$$

This solution represents two spherically symmetric waves—one traveling outward and one traveling inward. The leading factor of $\frac{1}{r}$ tells us how the amplitudes of the waves diminish as they progress.

We can see that $\frac{1}{r}$ is a physically reasonable rate of decay. Consider just the outward traveling wave:

$$\psi^{out} = \frac{1}{r}f(r - ct). \tag{4.53}$$

ψ is virtually always proportional to a quantity such as displacement, strain, stress, electromagnetic field strength, etc., and thus ψ^2 is usually a measure of energy density. The energy in the wavefront crossing a sphere of radius r at time t is then

$$4\pi r^2(\psi^2) = 4\pi f^2(r - ct) \tag{4.54}$$

the size of which is independent of radius (other than as a "phase" argument). Thus energy is conserved by the $\frac{1}{r}$ dependence.

4.3.3 The Sommerfeld Radiation Condition

The other spherically symmetric solution,

$$\psi^{in} = \frac{1}{r}g(R + ct) \tag{4.55}$$

is a wave traveling inward toward the origin. If the problem we were studying involved a source at the origin and if the medium was homogeneous and infinite, we might discard the inward-bound solution on the grounds that energy must flow outward from the source to infinity and not the other way around. This requirement is formally called the "Sommerfeld radiation condition" and is frequently applied in wave propagation problems.

The Sommerfeld condition applies equally in non-spherically symmetric problems; in those cases the separation into incoming and outgoing radiation may be a little more complicated. It is important to remember that the radiation condition only applies to radiation which is truly coming in from infinity; we cannot apply it inside of a region where incoming energy might legitimately appear as a result of distant scattering (unless, of course, we wish to neglect that scattering).

A rigorous statement of the Sommerfeld condition is this: let p be a solution of the Helmholtz equation outside of some finite sphere. Then it is required for a finite constant K that

$$|Rp| \; < \; K$$
$$R(\partial_R + ik)p \; \to \; 0 \tag{4.56}$$

uniformly with respect to direction as $R \to \infty$, where R is the radius of the sphere. In many circumstances it is possible to show *a priori* that solutions of the Helmholtz equation automatically satisfy this condition. So it is not too risky to assume that it applies.

4.3.4 Harmonic Waves

At a fixed point in space $\mathbf{r}_0$ a wave disturbance ψ is a function only of time

$$\psi(\mathbf{r}_0, t) = \Psi(t). \tag{4.57}$$

Because they can be used to synthesize completely general disturbances, an especially interesting case is when Ψ is periodic:

$$\Psi(t) = P\cos(\omega t). \tag{4.58}$$

Here P is the amplitude[1] and ωt is the phase. The number of vibrations per second is the frequency $f = \omega/2\pi$. The frequency is also the reciprocal of the period $f = 1/T$; you can readily verify that Ψ remains unchanged when $t \to t+T$.

The most useful harmonic wave is the harmonic plane wave:

$$\psi(\mathbf{r}, t) = P\cos[\mathbf{k} \cdot \mathbf{r} - \omega t] \tag{4.59}$$

The set of harmonic plane waves for all frequencies, ω, is complete. Thus any traveling wave can be built up out of a sum of plane waves. For more details on this point see [19].

Calculations with plane waves are simplified if we use complex notation. Instead of Equation (4.59) we write

---

[1] If we allow the amplitude P to be a function of space, then the surfaces of constant amplitude no longer necessarily coincide with the surfaces of constant phase. In that case the wave is called *inhomogeneous*.

$$\psi(\mathbf{r}, t) = \Re\left[P \exp\left(\mathbf{k} \cdot \mathbf{r} - \omega t\right)\right]. \tag{4.60}$$

As long as all the calculatons involved are linear, we can drop the $\Re$, keeping in mind that it's the real part of the final expression that we want. Nonlinear operations do come up however; for example, when computing the energy density we must square the displacement. In such cases one must take the real part first before performing the nonlinear operation.

4.4 Reflection

The imposition of boundary conditions at various surfaces in the medium generally gives rise to additional reflected wave fields. (In elasticity, boundaries also act, in general, to convert one type of wave into another upon reflection.) Reflected energy plays an essential role in many important wave propagation phenomena.

We will here examine only a simple case. Suppose we have a semi-infinite medium ending at $x = 0$. At the boundary surface, $x = 0$, we impose the boundary

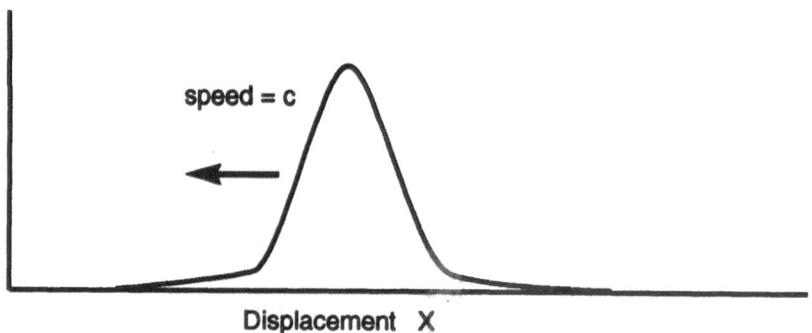

Fig. 4.2. A 1D pulse propagating towards the origin at speed c.

condition $\psi = 0$. Finally let

$$\psi^+ = f(x + ct) \tag{4.61}$$

be an incoming wave (that is, coming from the right, as in Figure 4.2). What is the solution to the wave equation which satisfies the boundary conditions?

We know that the general form of the solution of the 1D wave equation is

$$\psi(x, t) = f(x + ct) + g(x - ct), \tag{4.62}$$

where f is the known incoming wave and g is the (unknown) outgoing. The boundary condition has the form

$$\psi(0,t) = 0 \tag{4.63}$$

or

$$f(ct) + g(-ct) = 0 \tag{4.64}$$
$$g(-ct) = -f(ct). \tag{4.65}$$

So,

$$\psi(x,t) = f(x+ct) - f(ct-x) \tag{4.66}$$

The first term is just energy traveling to the left; the second is energy traveling to the right. Notice the reflected wave has the same form as the incident wave but is of opposite sign; obviously that is because the boundary condition we have applied requires the two waves to exactly cancel at $x = 0$. The argument of $f(x)$ is also reversed in sign in the reflected wave; this is because the spatial shape of the reflected wave is the reverse of that of the incident wave.

We have illustrated these ideas with a simple numerical example in Figure 4.3, which shows an incoming Gaussian pulse bouncing off a perfectly reflecting boundary at $t = 5$.

4.5 Phase Velocity, Group Velocity, and Dispersion

The next topic is somewhat outside of the subject of simple wave solutions to the scalar wave equation. It is, however, one that we must deal with and now is as good a point as any.

Suppose that we have a system governed by a wave equation which differs from the object of our current fascination in a rather strange way. We suppose that this system still supports plane harmonic waves of the form

$$\psi = e^{i(\omega t - \mathbf{k}\cdot\mathbf{r})} \tag{4.67}$$

where $\mathbf{k}$ is any vector but now we have

$$\omega = f(\mathbf{k}) \tag{4.68}$$

where $f(\mathbf{k})$ is some as yet unspecified function. In our previous case,

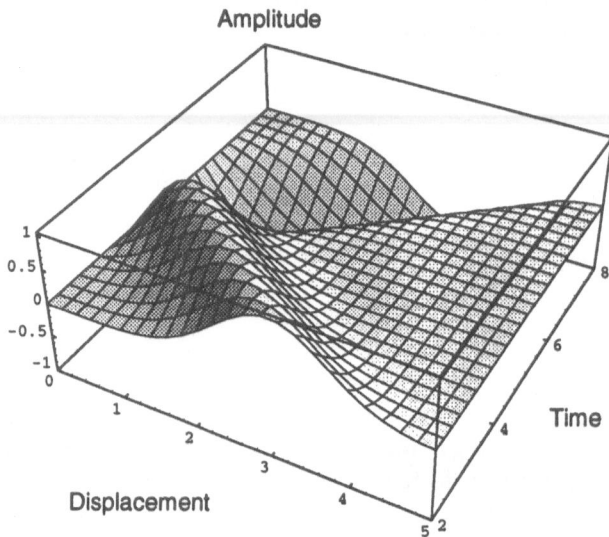

Fig. 4.3. A pulse incoming from the positive x direction and traveling with constant unit speed strikes a perfectly reflecting barrier at the origin at $t = 5$.

$$\omega = c|\mathbf{k}|. \tag{4.69}$$

We want to study the properties of wave motion in this system. The dependence of ω on $\mathbf{k}$ is called "dispersion" for reasons that will emerge.

This unusual behavior is not at all unusual and is more than just a fanciful assumption. Many types of wave propagation are noticeably dispersive including almost everything of interest to seismologists except, perhaps, body waves.

4.5.1 Model A

We will first construct a relatively uncomplicated model of the propagation of wave energy to see if we can infer some of the phenomena associated with dispersion. Consider the simultaneous propagation in the $+\hat{\mathbf{x}}$ direction of two plane waves. We have

$$\psi(x,t) = e^{i(\omega_1 t - k_1 x)} + e^{i(\omega_2 t - k_2 x)} \tag{4.70}$$

where

$$\omega_1 = f(k_1) \quad \omega_2 = f(k_2) \tag{4.71}$$

(For simplicity we have assumed ω depends only upon $|\mathbf{k}|$.) Now take

$$\begin{aligned} k_1 &= k - \epsilon \\ k_2 &= k + \epsilon \end{aligned} \qquad \text{where } \epsilon \ll k. \tag{4.72}$$

Then

$$\begin{aligned} \omega_1 &= \omega - \epsilon U \\ \omega_2 &= \omega + \epsilon U \end{aligned} \qquad \text{where} \begin{cases} \omega &= f(k) \\ U &= \partial_k f. \end{cases} \tag{4.73}$$

So

$$\psi = e^{i(\omega t - kx)} \left(e^{i\epsilon(-Ut+x)} + e^{i\epsilon(Ut-x)} \right) \tag{4.74}$$

$$\psi = 2e^{i(\omega t - kx)} \cos\{\epsilon(x - Ut)\} \tag{4.75}$$

The first factor in ψ is just the mean plane wave wiggling off to infinity. Because of interference between the two components, however, the amplitude of the wave is multiplied by a second factor which varies much more slowly in space (since $\epsilon \ll k$). Schematically, this is shown in Figure 4.4, which is a modulated version of the mean plane wave.

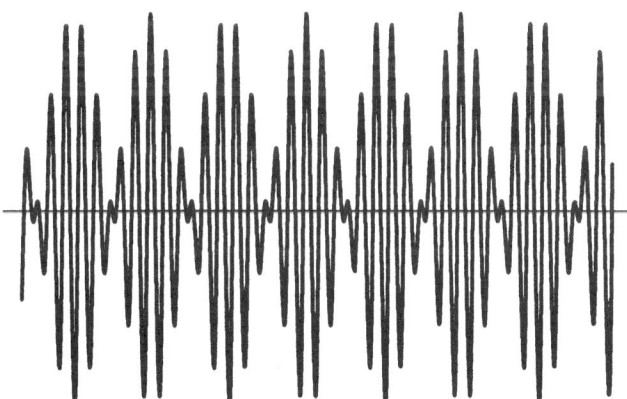

Fig. 4.4. Modulated sinusoid.

The zero-crossings of the composite wave travel with the speed

$$C = \frac{\omega}{k} \tag{4.76}$$

whether the wave is modulated or not. This quantity is called the "phase velocity." The modulation peaks and troughs, however, travel with the speed

$$U = \frac{\partial \omega}{\partial k} \tag{4.77}$$

which is called the "group velocity." The modulation envelope controls the local energy density $(\psi\bar{\psi})$ of the wave, where the bar denotes complex conjugation. Hence we associate group velocity with the speed of energy transport. Notice that both C and U depend only upon $\omega(k)$ and not upon ϵ, the wavenumber separation of the two interfering waves. The wavenumber separation, ϵ, does not control the modulation wavelength.

Observe that when $\omega = c/k$ then

$$C = U = c \tag{4.78}$$

In this case both phase and group velocity are equal to the velocity appearing in the wave equation. This condition is true for elastic waves propagating in an unbounded homogeneous medium.

This example serves more as a demonstration than a proof. The result we derived indicates a sinusoidal modulation envelope which itself extends infinitely far in both directions, and we might fairly wonder if our interpretation of the results is unfounded or at least suspect.

Such suspicion, while commendable, would be in error. The trouble with this demonstration lies in our use of only a finite number of interfering waves; with a finite set (two in this case) it is impossible to construct a spatially limited interference product.

For a more precise statement, we now present Model B. If the following is too involved for you, take heart in the fact that the answers are virtually the same as given above. Model A is all that you really need.

4.5.2 Model B

Construct a spatially bounded wave using a wavenumber integral over K, the 3-space of wavenumber vectors:

$$\psi(\mathbf{r}, t) = \int_K \phi(\mathbf{k}) e^{i(\omega t - \mathbf{k} \cdot \mathbf{r})} dv_k \tag{4.79}$$

where $\omega = f(\mathbf{k})$. All we are doing is adding up a continuum of plane waves; since each of these satisfies the wave equation (by hypothesis), so does the sum.

In particular, we take $\phi(\mathbf{k})$ to be a three-dimensional Gaussian distribution centered about $\mathbf{k}_0$,

$$\phi(\mathbf{k}) = (2\pi a^2)^{\frac{3}{4}} e^{-a^2(\mathbf{k}-\mathbf{k}_0)^2/4} \tag{4.80}$$

$$= N e^{-a^2(\mathbf{k}-\mathbf{k}_0)^2/4}, \quad \text{for short.} \tag{4.81}$$

We have selected a Gaussian distribution because it is band-limited in a practical sense, and because its inverse transform is spatially limited (the inverse is also Gaussian) in the same sense. So,

$$\psi(\mathbf{r}, t) = N \int_K e^{-a^2(\mathbf{k}-\mathbf{k}_0)^2/4} e^{i(\omega t - \mathbf{k}\cdot\mathbf{r})} dv_k. \tag{4.82}$$

Since $\phi(\mathbf{k})$ is effectively constrained to a region around $\mathbf{k}_0$, the size of which we control by selecting a, let us linearize around $\mathbf{k}_0$:

$$\begin{matrix} \mathbf{k} & = & \mathbf{k}_0 + \delta \\ \omega(\mathbf{k}) & = & \omega_0 + \delta \cdot \mathbf{U} \end{matrix} \quad \text{where} \quad \left\{ \begin{matrix} \omega_0 & = & \omega(\mathbf{k}_0) \\ \mathbf{U} & = & \nabla_\mathbf{k}\omega(\mathbf{k}) \end{matrix} \right. \tag{4.83}$$

$$\psi(\mathbf{r}, t) = N \int_K e^{\left(-a^2\delta\cdot\delta/4 - i\mathbf{k}_0\cdot\mathbf{r} - i\delta\cdot\mathbf{r} + i\omega_0 t + i\delta\cdot\mathbf{U}t\right)} dv_\delta \tag{4.84}$$

$$\psi(\mathbf{r}, t) = N e^{i(\omega_0 t - \mathbf{k}_0\cdot\mathbf{r})} \int_K e^{\left(-a^2\delta\cdot\delta/4 - i\delta\cdot\mathbf{r} + i\delta\cdot\mathbf{U}t\right)} dv_\delta \tag{4.85}$$

$$\psi(\mathbf{r}, t) = N e^{i(\omega_0 t - \mathbf{k}_0\cdot\mathbf{r})} \int_K e^{\left(-a^2\delta\cdot\delta/4 + i\delta\cdot\mathbf{P}\right)} dv_\delta \tag{4.86}$$

where $\mathbf{P} = \mathbf{U}t - \mathbf{r}$ \hfill (4.87)

Call the integral $H(\mathbf{P})$. Convert it to spherical polar coordinates (δ, θ, ϕ) by taking $\mathbf{P}$ as the pole and let $\alpha = \cos\theta$.

$$H(\mathbf{P}) = 2\pi \int_0^\infty \delta^2 d\delta \int_{-1}^{+1} e^{-a^2\delta^2/4 + i\delta P\alpha} d\alpha \tag{4.88}$$

$$P = |\mathbf{P}|. \tag{4.89}$$

Since

$$\int_{-1}^{+1} e^{i\delta P\alpha} d\alpha = \frac{2}{\delta P} \sin(\delta P) \tag{4.90}$$

$$H(\mathbf{P}) = \frac{4\pi}{P} \int_0^\infty \delta \sin(\delta P) e^{-a^2\delta^2/4} d\delta \tag{4.91}$$

$$H(\mathbf{P}) = \frac{4\pi}{P} \int_{-\infty}^\infty \frac{\delta}{2i} e^{i\delta P} e^{-a^2\delta^2/4} d\delta. \tag{4.92}$$

Substitute

$$q = \frac{\delta a}{2} + \frac{iP}{a} \tag{4.93}$$

$$\delta = \frac{2}{a}\left(q - \frac{iP}{a}\right) \tag{4.94}$$

to get

$$H(\mathbf{P}) = \frac{4\pi}{P}\frac{2}{a}e^{-P^2/a^2}\int_{-\infty}^{\infty}\left(q - \frac{iP}{a}\right)e^{-q^2}dq \tag{4.95}$$

$$H(\mathbf{P}) = \frac{-i8\pi}{a^2}\sqrt{\pi}e^{-P^2/a^2} \tag{4.96}$$

since

$$\int_{-\infty}^{\infty} qe^{-q^2}dq = 0 \tag{4.97}$$

So,

$$\psi(\mathbf{r},t) = \frac{-8\pi iN}{a^2}\sqrt{\pi}e^{i(\omega_0 t - \mathbf{k}_0\cdot\mathbf{r})}e^{-(Ut-r)^2/a^2} \tag{4.98}$$

This is the three-dimensional version of our earlier result. Note that $\mathbf{U} = \nabla_k f$ is a vector quantity and need not be parallel to $\mathbf{k}_0$.

4.5.3 Dispersion on a Lattice

This section may appear at first to be a little off the beaten path, but in fact it holds the key to a pervasive kind of error which will crop up when we consider finite difference methods of migration. When we attempt to solve the wave equation numerically by finite differences we immediately discover that any finite approximation to the continuum equations introduces an artificial dispersion identical to the dispersion just discussed. This is a result of the fact that a finite dimensional system has only finitely many degrees of freedom, and that any attempt to propagate energy of a wavelength smaller than the grid spacing will result in that energy being dispersed amongst the longer wavelength components of the solution. Since it is important to recognize that this intrinsic drawback to finite difference approximation is indistinguishable from true physical dispersion in a dynamical system, we introduce the concept here in terms of wave propagation on a 1D lattice, which we use as a finite dimensional approximation of a continuous string (i.e., a 1D medium). We will see that direct appeal to Newton's laws of motion gives rise to precisely the finite difference equations that we will study in detail later and that the dispersion relation follows readily from these discrete equations of motion. Further we will show that as the wavelength of the disturbance propagating on the lattice gets larger and larger, the disturbance no longer sees the granularity of the medium and propagates nondispersively.

The lattice will be populated with points of mass m, uniformly distributed along the line. The position of the ith particle will be x_i, whereas its equilibrium positions will be X_i. The displacement from equilibrium is

$$u_i \equiv x_i - X_i. \tag{4.99}$$

The first step is to consider small displacements from equilibrium. If we denote the equilibrium energy of the system by $V_0 \equiv V(X)$ and expand V in a Taylor series about X, retaining only terms to second order, we have

$$V = V_0 + \frac{1}{2} \sum_l \sum_{l'} V_{ll'} u_l u_{l'} \tag{4.100}$$

where the sums are over all lattice elements and where

$$V_{ll'} \equiv \left[\frac{\partial^2 V}{\partial x_l \partial x_{l'}} \right]_X . \tag{4.101}$$

We can reference the state of the system to its equilibrium by defining the interaction energy $U \equiv V - V_0$

$$U = \frac{1}{2} \sum_l \sum_{l'} V_{ll'} u_l u_{l'}. \tag{4.102}$$

On the other hand, the kinetic energy is

$$T = \frac{1}{2} m \sum_l \dot{u}_l^2. \tag{4.103}$$

Since rotations are irrelevant in 1D, the only conservation principles that we require are invariance of the interaction energy under changes of particle numbering and rigid translations. The particle numbering reflects both a choice of origin and a direction of increasing particle number; that is $l \to l + p$ and $l \to -l$. Thus we require that the energy satisfy

$$V_{ll'} = V_{l-l'} = V_{l'-l}. \tag{4.104}$$

This, together with the requirement that rigid translations $u_l = \delta$ not change the energy implies that

$$\left[\frac{1}{2}\sum_l \sum_{l'} V_{l-l'}\right] \delta^2 = 0,$$ (4.105)

which in turn implies that

$$\sum_{l'} V_{l-l'} = V_0 + \sum_{l' \neq l} V_{l-l'} = 0.$$ (4.106)

Since we already have convenient expressions for the kinetic and potential energies, the equations of motion follow directly from the Euler-Lagrange equations

$$\frac{d}{dt}\frac{\partial L}{\partial \dot{u}_l} - \frac{\partial L}{\partial u_l} = 0$$ (4.107)

where $L \equiv T - U$. Substituting in the computed expressions for the kinetic and potential energies, one readily verifies that the equations of motion are

$$m\ddot{u}_l = \sum_{l'} V_{ll'} u_{l'}.$$ (4.108)

We can put Equation (4.108) in a more transparent form by first putting $V_{ll'} = V_{l-l'}$ in Equation (4.108). Then using Equation (4.105), which allows us to add and subtract $\sum_{l'} V_{ll'} u_l$, and changing the variable of summation to p we are left with

$$m\ddot{u}_l = \sum_{p=1}^{\infty} V_p(u_{l+p} - 2u_l + u_{l-p}).$$ (4.109)

By a similar argument, one can show that the quadratic potential energy is

$$U = \frac{1}{2}\sum_{l=-\infty}^{+\infty}\sum_{p=1}^{\infty} V_p(u_{l+p} - u_l)^2.$$ (4.110)

The interpretation of equation Equation (4.109) is very simple, since it represents for any p, say $p = i$, the "elastic" force on the lth mass due to mass points symmetrically placed at $l+p$ and $l-p$ and with spring constants V_p. It also gives us a very simple way of quantifying the range of the molecular interactions. Retaining only terms involving V_1 yields a model with nearest neighbor interactions. Going to $p = 2$ gives nearest-plus-next-nearest neighbor interactions, and so on.

The plane wave dispersion relation for the perfect 1D lattice is obtained by substituting the trial-solution

$$u_l \equiv U e^{i(kal - \omega t)} \tag{4.111}$$

into Equation (4.109), where a is the lattice spacing. The calculation is straight-forward and Equation (4.109) implies that

$$-m\omega^2 u_l = -2u_l \sum_{p=1}^{\infty} V_p [1 - \cos(kap)]. \tag{4.112}$$

This equation can be simplified somewhat by using the trigonometric identity $\cos(2A) = 1 - 2\sin^2(A)$. Thus we arrive at the dispersion relation

$$m\omega^2 = 4 \sum_{p=1}^{\infty} V_p [\sin^2(kap/2)]. \tag{4.113}$$

The simplest case to handle is that of nearest neighbor interactions $V_p = \delta_{p1}\kappa$, where δ_{ij} is the Kronecker delta function. In which case

$$\omega = 2\sqrt{\frac{\kappa}{m}} \sin(ka/2). \tag{4.114}$$

Owing to periodicity and symmetry, it suffices to consider the range of wavenumbers $0 \le k \le \frac{\pi}{a}$, the so-called first Brillouin zone. The group velocity is

$$v_g \equiv \frac{d\omega}{dk} = \frac{a}{2}\omega_{max} \cos(ka/2), \tag{4.115}$$

where ω_{max} is simply $2\sqrt{\frac{\kappa}{m}}$. Thus the maximum speed with which information can propagate on the lattice is

$$v_{max} \equiv \max\, v_g = \frac{a\omega_{max}}{2} = a\sqrt{\frac{\kappa}{m}}. \tag{4.116}$$

Finally, for a finite lattice we must impose boundary conditions at the endpoints. Suppose we have $N+1$ oscillators running from $l = 0$ to $l = N$. The simplest case is if we fix the endpoints. In which case the jth normal mode is

$$u_l^j = C_l \sin\left(\frac{\pi j l}{N+1}\right) \cos(\omega_j t) j = 1, ..., N. \tag{4.117}$$

And its eigenfrequency is

$$\omega_j = 2\sqrt{\frac{\kappa}{m}} \sin\left(\frac{\pi j}{2(N+1)}\right). \tag{4.118}$$

It is easy to see that the total energy of each eigenmode is independent of time. A plot of some of these modes for a homogeneous lattice of 50 mass points is given in Figure 4.5. Notice especially that for the longer wavelengths, the modes are pure sinusoids, while for shorter wavelengths, the modes become modulated sinusoids. This is another manifestation of the dispersive effects of the discrete system.

The Continuum Equations. It is always useful to have the continuum limit of a given set of lattice equations. To arrive at the continuum equations for the nearest neighbor problem we simply replace u_l by $u(x_l)$ in the equations of motion Equation (4.109) (but with $V_p = \delta_{1p}\kappa$) and perform Taylor series expansions of the terms about x_l:

$$u_{l+1} \equiv u(x_l + a) \tag{4.119}$$

$$= u(x_l) + \frac{\partial u}{\partial x_l}a + \frac{1}{2}\frac{\partial^2 u}{\partial x_l^2}a^2 + \dots \tag{4.120}$$

Keeping only terms of second order or less, the $p = 1$ term of Equation (4.109) gives

$$m\frac{\partial^2 u}{\partial t^2} = \kappa a^2 \frac{\partial^2 u}{\partial x_l^2}. \tag{4.121}$$

The last step is to approximate a density function by averaging the point mass over adjacent lattice sites, $\rho = \frac{m}{a^3}$, in which case Equation (4.121) becomes

$$\rho\frac{\partial^2 u}{\partial t^2} = \frac{\kappa}{a}\frac{\partial^2 u}{\partial x_l^2}, \tag{4.122}$$

from which we can identify Young's modulus as $E \equiv \frac{\kappa}{a}$.

The continuum dispersion relation is likewise computed by substituting a plane wave into Equation (4.122), with the result

$$\omega = k\sqrt{\frac{E}{\rho}}. \tag{4.123}$$

And the continuum group velocity is simply

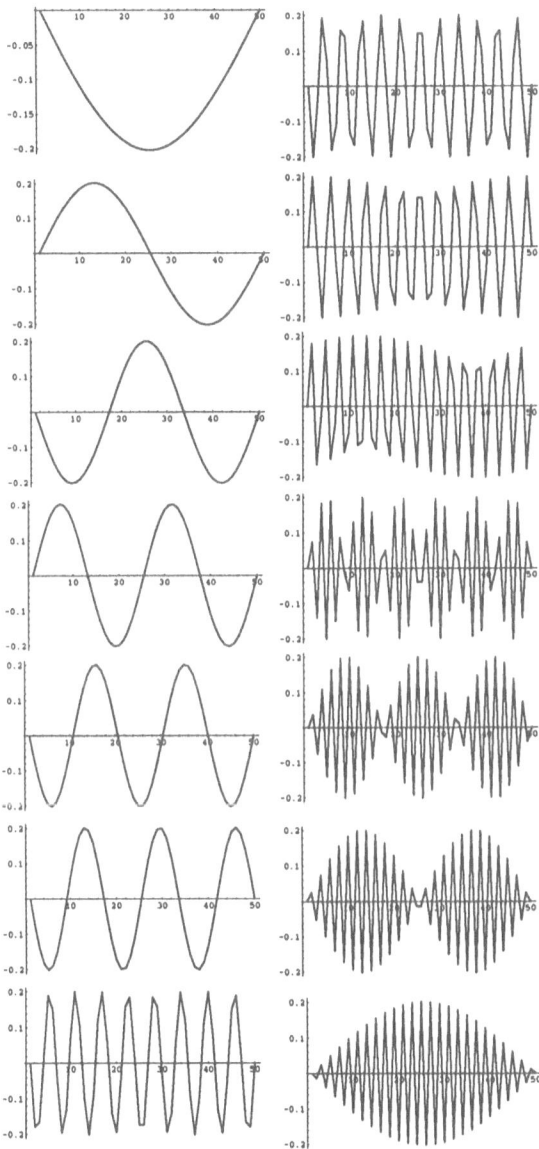

Fig. 4.5. A sample of the normal modes (free oscillations) of a homogeneous 50 point lattice with fixed ends. The lower frequency modes are purely sinusoidal; the higher frequency modes become modulated sinusoids as a result of the dispersive effects of this being a discrete system.

$$\frac{d\omega}{dk} = \sqrt{\frac{E}{\rho}} = a\sqrt{\frac{\kappa}{m}}, \tag{4.124}$$

which agrees with the lattice theoretic result of Equation (4.115) at small wavenumbers. This is precisely what one would expect since at small wavenumbers or large wavelengths a propagating wave does not see the granularity of the lattice. The departure of the lattice dispersion relation from the continuum is shown in Figure 4.6.

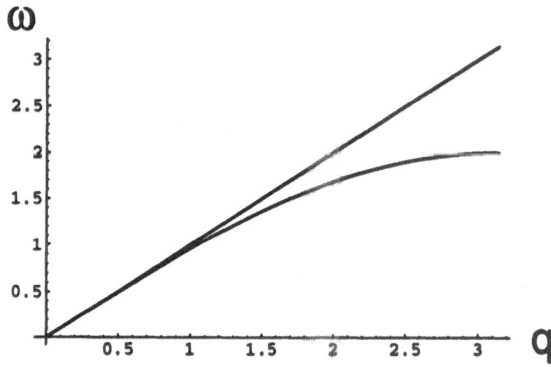

Fig. 4.6. Discrepancy between the dispersion-free string and the finite dimensional dynamical system used to approximate it on a computer. The lattice spacing was chosen to be unity.

4.5.4 Summary of Dispersion

Let us briefly summarize the new effects associated with dispersive wave propagation:

- If the medium is dispersive, the phase velocity is different for each frequency component. As a result, different components of the wave move at different speeds and therefore tend to change phase relative to one another.

- In a dispersive medium the speed with which energy flows may differ greatly from the phase velocity.

- Although we haven't talked about attenuation, in a dissipative medium, a pulse will be attenuated as it travels with or without dispersion, depending on whether the dissipative effects are sensitive to frequency.

4.6 A Quantum Mechanical Slant on Waves

What do we mean when we say that a pulse arrives at a certain time? Following [16], let us consider a probabilistic approach to this question inspired by quantum mechanics. The eikonal equation

$$(\nabla \Phi(\mathbf{r}))^2 - s^2(\mathbf{r}) = 0 \tag{4.125}$$

describes the high frequency propagation of seismic waves; here Φ is the wavefront and s is the slowness. In seismic applications we almost always treat the depth z as a preferred coordinate since our media are vertically stratified for the most part. So let's think of z as being more of a parameter of motion and rewrite the eikonal equation as

$$\frac{\partial \Phi}{\partial z} + H(x, y, ; p_x, p_y; z) = 0 \tag{4.126}$$

where $p_x = \partial \Phi / \partial x$ and $p_y = \partial \Phi / \partial y$ and H is given by

$$H(x, y, ; p_x, p_y; z) = -\left[s^2(x, y, z) - p_x^2 - p_y^2 \right]^{1/2}. \tag{4.127}$$

H is called the Hamiltonian and Equation (4.126) is an example of what is known in physics as a Hamilton-Jacobi equation. A good reference for this sort of thing is Chapter 9 of Goldstein's book *Classical Mechanics* [25].

Since the Hamiltonian has been obtained from the eikonal equation, it is clear that it gives a representation in terms of rays. The classical mechanical analog of rays is particles. Now when we make the transition from the classical mechanics of particles to the quantum mechanics of wave packets, we interpret the generalized momenta p_x and p_y in the Hamiltonian as operators:

$$p_x = -i \frac{h}{2\pi} \frac{\partial}{\partial x} \tag{4.128}$$

$$p_y = -i \frac{h}{2\pi} \frac{\partial}{\partial y} \tag{4.129}$$

$$H = -i \frac{h}{2\pi} \frac{\partial}{\partial z} \tag{4.130}$$

where h is Planck's constant, except that in the mechanics problem we would be thinking in terms of time as the parameter, not depth. These operators are interpreted strictly in terms of their action of some wave function Ψ so that, for example, when we write a commutation relation such as

$$[p_x, x] = i \frac{h}{2\pi}$$

this is really shorthand for

$$[p_z, x]\Psi = \{p_z x\Psi - x p_z \Psi\}$$
$$= i\frac{h}{2\pi}\{\Psi - x\Psi' + x\Psi'\} = i\frac{h}{2\pi}\Psi.$$

In our case, we can identify Planck's constant with one over the frequency: as the frequency goes to infinity we must recover classical (i.e., ray-theoretic) physics. So we have

$$p_x = -i\frac{1}{\omega}\frac{\partial}{\partial x} \tag{4.131}$$

$$p_y = -i\frac{1}{\omega}\frac{\partial}{\partial y} \tag{4.132}$$

$$H = -i\frac{1}{\omega}\frac{\partial}{\partial z}. \tag{4.133}$$

If we plug these operators into the definition of the Hamiltonian we arrive at none other than the Helmholtz equation:

$$\nabla^2\Psi + k^2\Psi = 0, \quad k = \omega s. \tag{4.134}$$

But now we're in business because we can interpret the solution of the wave equation probabilistically as in quantum mechanics. For example, at a fixed point in space we can normalize the wave function Ψ as

$$1 = \int \Psi^*\Psi \, dt.$$

Then the function $|\Psi|^2$ can be interpreted as the probability density of the arrival of a wavefront at a time t. This allows us to make a rigorous, wave-theoretic concept of arrival time since now we can say that the expectation value of the arrival time is

$$\langle t \rangle = \int \Psi^* t \Psi \, dt. \tag{4.135}$$

So naturally, we want to identify the width of a pulse with its "standard deviation"

$$\delta^2 = \int \Psi^*(t - \langle t \rangle)^2 \Psi \, dt. \tag{4.136}$$

Let us therefore define the arrival time of a pulse as:

$$t_a = \langle t \rangle - \delta. \tag{4.137}$$

Well, it looks plausible on paper, but does it work? In Figure 4.7, we show some synthetic "wavelets". These are just the first lobe of a Ricker wavelet, perturbed by uniformly distributed noise. On the right of each plot is the average time and arrival time as defined above. We will leave the plausibility of these numbers to your judgement. Of course, there is nothing magical about one standard deviation. We could use a different measure, say three standard deviations, on the grounds that this would contain almost all of the energy of the pulse. The main point is that Equation (4.137) is a quantitative and easily computed measure.

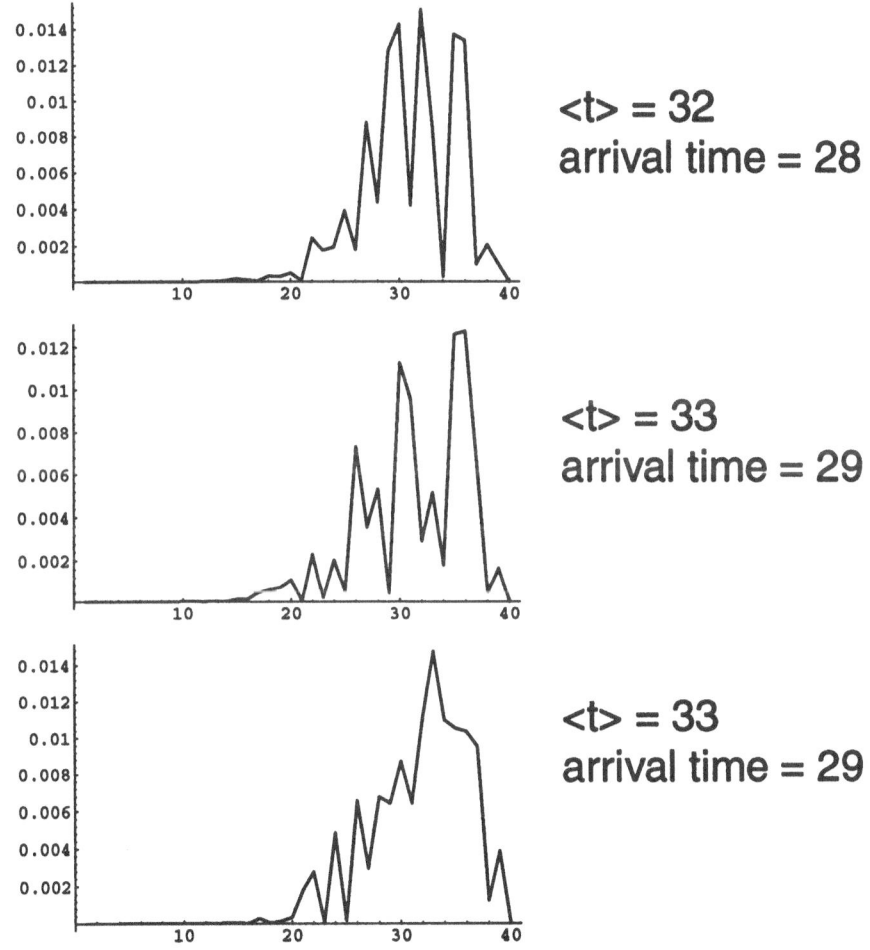

Fig. 4.7. Synthetic Ricker lobes with noise added.

As usual, real data are somewhat more problematic. In Figure 4.8, we show a trace taken from the Conoco crosshole data. We looked around for one that looked especially clean. On the right of the figure you see a zoomed-in view around the "first break". Hard to tell from this view so we just squared the trace amplitude. Figure 4.9 shows a further zoom of the squared trace. Here the rise in amplitude associated with the arrival is relatively easy to pick out. The "quantum mechan-ical" calculation gives an arrival time of around 7 in this particular picture. For more details of this "quantum mechanical" approach see [16].

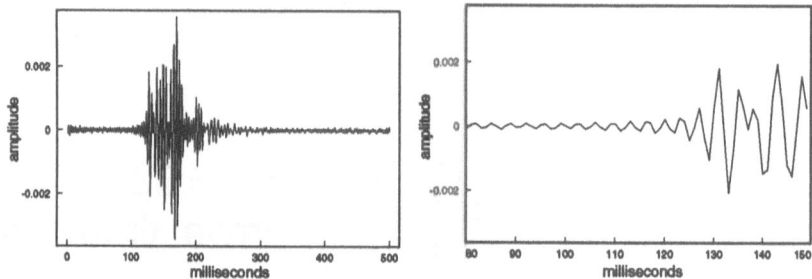

Fig. 4.8. Left: A trace taken from the Conoco crosshole data set. Right: An enlarged view of the portion containing the first arrival.

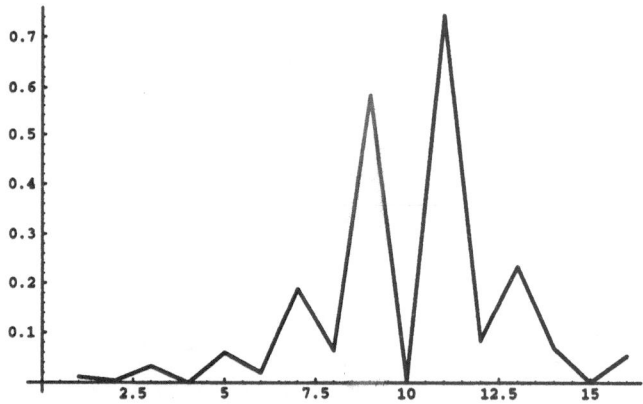

Fig. 4.9. Enlarged view of the squared trace. Don't pay attention to the x-axis labeling. The y-axis represents the normalized amplitude.

Exercises

4.1 What is the physical significance of the fact that shear waves are dilation-free ($\nabla \cdot \mathbf{s} = 0$)?

4.2 Do shear waves ever propagate faster than compressional waves? If not, why not?

4.3 Derive the wave equation in spherical coordinates (r, θ, ϕ). Then specialize to the case of a spherically symmetric medium.

4.4 In Chapter 2 you computed the Fourier transform of a Gaussian. Imagine that this Gaussian is the initial state of a pulse traveling down the x−axis: $u(x,0)$. Compare the width of $|u(x,0)|^2$ (call this Δx) with the width of its Fourier transform squared: $|A(k)|^2$ (call this Δk). What conclusions can you draw about how Δx and Δk are related?

4.5 Show that if

$$\psi(\mathbf{r}, t) = f(\hat{\mathbf{n}} \cdot \mathbf{r} - ct)$$

then

$$\nabla \psi = \hat{\mathbf{n}} f'$$
$$\nabla \cdot (\nabla \psi) = \hat{\mathbf{n}} \cdot \hat{\mathbf{n}} f'' = f''$$

and

$$\partial_t \psi = -cf'$$
$$\partial_t^2 \psi = c^2 f''$$

4.6 Show that the peaks and troughs of the modulated sinusoid propagate with the speed $U = \partial \omega / \partial k$.

5. Ray Theory

5.1 The Eikonal Equation

We will concentrate on the canonical scalar wave equation:

$$\partial_t^2 \psi = c^2 \nabla^2 \psi \tag{5.1}$$

where ψ is some potential. We Fourier-transform from time, t, to angular frequency, ω, so that

$$\partial_t \rightarrow i\omega \tag{5.2}$$

and we have

$$-\omega^2 \psi = c^2 \nabla^2 \psi. \tag{5.3}$$

Plane-wave solutions for (transformed) ψ are of the general form

$$\psi = Pe^{i\Phi} \tag{5.4}$$
$$\Phi = \mathbf{k} \cdot \mathbf{r} \tag{5.5}$$
$$|k| = \left|\frac{\omega}{c}\right| \tag{5.6}$$

for the case where c, the wave speed, is everywhere constant.

Now suppose that c varies with position but slowly enough that we can use the wave equation

$$-\omega^2 \psi = c^2(\mathbf{r}) \nabla^2 \psi. \tag{5.7}$$

This is called the Helmholtz equation. Basically such a supposition amounts to neglecting the coupling terms (involving things like ∇c or $\nabla \lambda$) which most exact wave equations possess. In general, as frequency increases we expect this approximation to become increasingly valid. Define the index of refraction, γ, with respect to some (arbitrary) reference speed c_0 by

$$\gamma(\mathbf{r}) = \frac{c_0}{c(\mathbf{r})}. \tag{5.8}$$

Now we will investigate the *possibility* that ψ may have (possibly approximate) solutions of the form

$$\psi = P(\mathbf{r})e^{i\Phi(\mathbf{r})} \tag{5.9}$$

where $P(\mathbf{r})$ is a real, "slowly-varying" wave amplitude and $\Phi(\mathbf{r})$ is a real generalized phase. (Remember that we are in the frequency domain; we regard ω as fixed, of course.)

We plug our assumed form of ψ into the wave equation and see what comes out:

$$\nabla\psi = (\nabla P)e^{i\Phi} + (i\nabla\Phi)(Pe^{i\Phi}) \tag{5.10}$$

$$\nabla^2\psi = \nabla \cdot \nabla\psi \tag{5.11}$$

$$= (\nabla^2 P)e^{i\Phi} + 2i(\nabla P \cdot \nabla\Phi)e^{i\Phi} \tag{5.12}$$

$$+ i\nabla^2\Phi(Pe^{i\Phi}) - (\nabla\Phi \cdot \nabla\Phi)(Pe^{i\Phi}) \tag{5.13}$$

so,

$$-\omega^2 Pe^{i\Phi} = c^2(\nabla^2 P - (\nabla\Phi \cdot \nabla\Phi)P)e^{i\Phi} \tag{5.14}$$

$$+ ic^2(2\nabla P \cdot \nabla\Phi + P\nabla^2\Phi)e^{i\Phi}. \tag{5.15}$$

If P, Φ satisfy this equation, then ψ as specified above satisfies the wave equation (to be more precise, the approximate wave equation when c is a slowly-varying function of position). Equating real and imaginary parts individually we get

$$-\omega^2 P = c^2\nabla^2 P - c^2 P\nabla\Phi \cdot \nabla\Phi \tag{5.16}$$

or

$$(\nabla\Phi)^2 = \frac{\omega^2}{c^2} + \frac{\nabla^2 P}{P} \tag{5.17}$$

for the real part and

$$\nabla^2\Phi + 2\nabla \log P \cdot \nabla\Phi = 0 \tag{5.18}$$

for the imaginary.

We now suppose that

$$(\nabla\Phi)^2 \gg \frac{\nabla^2 P}{P} \tag{5.19}$$

to such an extent that the second term in Equation (5.17) is negligible. When c is a constant in space the P is also a constant and $\nabla^2 P$ is exactly zero while $(\nabla \Phi)^2$ is exactly equal to ω^2/c^2. When c is no longer constant but still reasonably smooth, neglecting $\nabla^2 P/P$ is, of course, an approximation. We justify this by claiming that at sufficiently high frequencies the wavenumber of the solution is much, much greater than the rate at which wave amplitude, or energy, varies in space. There are two important features of this approximation:

1. The approximation in general improves as frequency increases. In the limit of unbounded frequency it becomes arbitrarily good. Because of this feature, we sometimes call this approach "asymptotic ray theory."

2. At any finite frequency there are always cases in which ray theory fails. We must always be careful in exploiting this technique.

The equation

$$(\nabla \Phi)^2 = \frac{\omega^2}{c^2} \tag{5.20}$$

is called the "eikonal" equation, for the Greek word $\epsilon\iota\kappa\tilde{\omega}\nu$ meaning "image." The solution, Φ, is sometimes called the "eikonal." The solutions, Φ, to this non-linear partial differential equation are the phase surfaces associated with high-frequency wave propagation through a medium with, at worst, a slowly-varying wave speed.

Note that $\nabla \Phi$ is essentially the local wave-number. To see this, we neglect possible variation in P and expand

$$\begin{aligned}
\psi(\mathbf{r} + \delta) &= P \exp(i\{\Phi(\mathbf{r} + \delta)\}) \tag{5.21} \\
&= P \exp(i\{\Phi(\mathbf{r}) + \delta \cdot \nabla \Phi(\mathbf{r})\}). \tag{5.22}
\end{aligned}$$

Basically, the eikonal equation is telling us how the local solution wavenumber is related to the local (and only the local) material properties.

Equation (5.18) is called the transport equation, since once we solve the eikonal equation for the phase, we can compute the slowly varying change in amplitude, which is due to geometrical spreading.[1]

---

[1] In fact Equation (5.9) is the first in an infinite series of transport equations obtained by making the more general *ansatz*:

$$\psi = e^{i\Phi(\mathbf{r})}(p_1 + p_2/k_0 + p_3/k_0^2 + \cdots)$$

It is not obvious that this expression is legitimate either as a convergent or an asymptotic series. Nevertheless, experience is overwhelmingly in its favor. For more details see [32].

It's worth pointing out that we could have proceeded in an entirely equivalent derivation of the eikonal equation by defining the phase in Equation (5.9) relative to the frequency (i.e., use use $e^{ik\Phi'(\mathbf{r})}$ instead). Plugging this *ansatz* into the Helmholtz equation, we would have seen several terms with a k in front of them. Then in the limit that the frequency goes to infinity, only these terms would have been non-negligible.

5.2 The Differential Equations of Rays

Consider the surfaces associated with constant values of the eikonal. The normal, n̂, to these surfaces are given by

$$\hat{\mathbf{n}} = \frac{\nabla \Phi}{|\nabla \Phi|} = \frac{\nabla \Phi}{k_0 \gamma} \text{ where } k_0 = \frac{\omega}{c_0}. \tag{5.23}$$

We define a trajectory by the locus of points, $\mathbf{r}(\sigma)$, where σ is arclength, such that

$$d_\sigma \mathbf{r} = \hat{\mathbf{n}}(\mathbf{r}). \tag{5.24}$$

We call this trajectory a "ray." We will see later why we attach so much significance to the ray. Right now, we wish to find a way to construct the ray without having to directly solve the eikonal equation. We do this as follows:

$$d_\sigma(\mathbf{r}) = \hat{\mathbf{n}} = \frac{\nabla \Phi}{k_0 \gamma} \tag{5.25}$$

$$\gamma d_\sigma \mathbf{r} = \frac{1}{k_0} \nabla \Phi \tag{5.26}$$

$$d_\sigma(\gamma d_\sigma \mathbf{r}) = \frac{1}{k_0} d_\sigma(\nabla \Phi). \tag{5.27}$$

Since

$$d_\sigma = \hat{\mathbf{n}} \cdot \nabla \tag{5.28}$$

$$d_\sigma(\gamma d_\sigma \mathbf{r}) = \frac{1}{k_0} \hat{\mathbf{n}} \cdot \nabla(\nabla \Phi) \tag{5.29}$$

$$= \frac{1}{\gamma k_0^2} \nabla \Phi \cdot \nabla(\nabla \Phi) \tag{5.30}$$

$$= \frac{1}{\gamma k_0^2} \frac{\nabla(\nabla \Phi \cdot \nabla \Phi)}{2} \quad \text{(why?)} \tag{5.31}$$

$$= \frac{1}{2\gamma k_0^2} \nabla(\gamma^2 k_0^2) \tag{5.32}$$

so

$$d_\sigma(\gamma d_\sigma \mathbf{r}) = \nabla\gamma \qquad (5.33)$$

or

$$d_\sigma(\gamma \hat{\mathbf{n}}) = \nabla\gamma. \qquad (5.34)$$

This equation describes the precise manner in which a ray, as we have defined it, twists and turns in response to the spatial variation of the wave speed, c. Imagine starting at some initial point, $\mathbf{r}(0)$, and in some initial direction, $\hat{\mathbf{n}}(0)$. Then in the presence of some particular $\gamma(\mathbf{r})$, this equation tells us how to take little local steps along the unique ray specified by $\mathbf{r}(0)$, $\hat{\mathbf{n}}(0)$.

From Equation (5.34) it is obvious that if the index of refraction (or velocity, or slowness) is constant, then the equation of a ray is just $d_\sigma^2 \mathbf{r} = 0$, which is the equation of a straight line. Notice that once the rays have been found, the eikonal can be evaluated since its values at two points on a ray differ by $\int \gamma d\sigma$. Finally, we note that the curvature of a ray is defined to be $d_\sigma \mathbf{r}$. But this must point in a direction normal to the ray and in the direction of increasing curvature. If we define a new unit normal vector along the ray path $\hat{\nu}$, then the radius of curvature is defined by the equation

$$d_\sigma \mathbf{r} = \frac{1}{\rho}\hat{\nu}. \qquad (5.35)$$

Using the ray equation, Equation (5.34), and the eikonal equation it can be shown that

$$\frac{1}{\rho} = \hat{\nu}\nabla\log\gamma \qquad (5.36)$$

which shows that rays bend towards the region of higher slowness (or index of refraction in optics). Also notice that Equation (5.34) is precisely the equation of a particle in a gravitational potential $-\gamma$. So in a deep sense, geometrical optics is equivalent to the classical mechanics of particles–Fermat's principle of least time being the exact analog of Hamilton's principle of least action, about which more in the next section.

5.3 Fermat's Principle

We begin with Huygen's Principle which asserts that as a wave disturbance spreads through a medium, we may, at each instant, regard the points disturbed by the ray at that moment as a new set of radiating sources. To take a very simple example, consider a pulse from some source which at time t has spread outward to the surface ∂V: Huygen's Principle, as we have already seen, says that at any later time the pulse may be found by computing the signal from a set of sources appropriately distributed over the disturbed area contained by ∂V. In this form, Huygen's Principle is simply a statement of the obvious fact that we can regard the calculation for times later than some moment, t, as being the solution of an initial-value problem starting at t having initial displacements and velocities equal to those they actually had at t.

Huygen's Principle suggests a rather incoherent, jumbled progress of wave energy outward from the source. It implies that wave energy traverses every possible path available to it. We, on the other hand, hope to find some more orderly scheme of wave propagation.

Harmony among these divergent notions comes as follows: Consider a medium in which the wave speed strictly increases with depth (but "slowly" of course) and in which we have a source and a receiver, both at some common depth. In Figure 5.1

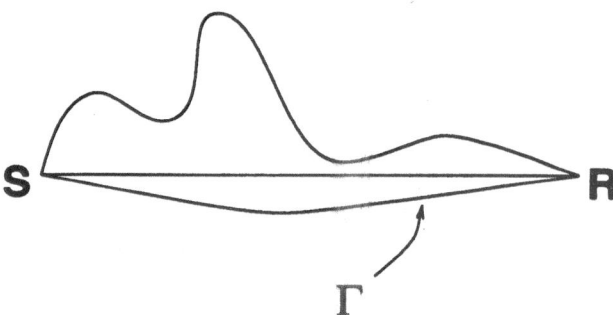

Fig. 5.1. Possible ray paths connecting the source and receiver.

we have drawn in three trial paths by which radiated energy might reach the receiver; one of them is somewhat fanciful. Any continuous piecewise differentiable curve is obviously a possible path. Because of the multitude of unrelated paths, we might expect, and it is so, that the signals arriving from the vast majority of them will interfere destructively to produce no net observable effect at the receiver.

Now contemplate the path with the distinguishing property that it is the path from source to receiver having the shortest time-of-flight (travel-time). In our particular case this path will not be a straight line from source to receiver but, because the wave speed increases downward, the minimum-time path will "dip" down somewhat into the higher velocity region. The path is sketched (qualitatively) as the lowermost path on the figure.

Let Γ symbolically denote this path and let η be a small path-variation which vanishes at either end. (We are playing a little fast and loose with algebra here, but it will work out all right.) Symbolically, the set of paths

$$\Gamma + \alpha\eta \qquad\qquad (5.37)$$

for small α, is a group of paths from source to receiver near to, and containing, the minimum-time path Γ. Since Γ is a *minimum*-time path, none of the nearby paths can be any faster than Γ and most must be slower. For any particular η, travel-time as a function of α must look like Figure 5.2 because the bottom

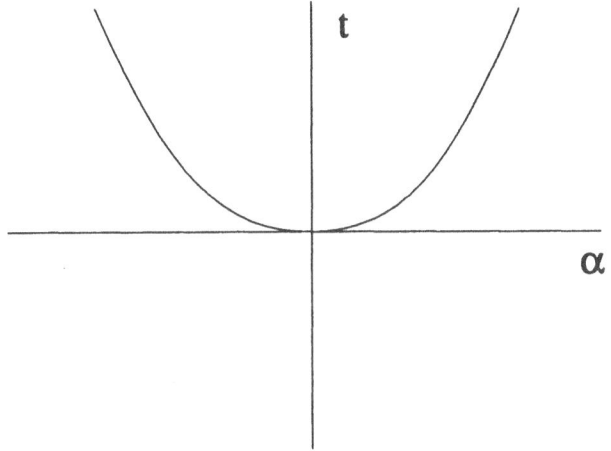

Fig. 5.2. The minimum time path is (by definition) an extremum of the travel time as a function of ray path.

of this curve is flat, as is always the case for a minimum; thus there must be a cluster of paths all having the same travel-time and which will therefore interfere constructively to produce a signal at the receiver. One of the most important properties of a minimum-time path, then, is that it delivers a significant amount of coherent energy at the far end.

Notice that this argument only really required that the minimum-time path be a *local* minimum. Between any pair of points in a given medium, there may be a number of minimum-time (in the local sense) paths and each should contribute an "arrival."

Fermat's Principle asserts that a *ray* is a miminum-time path through the medium. When we have shown that this is so, we shall have found out why ray theory is useful.

We begin by proving Lagrange's integral invariant. Since

$$\gamma \hat{\mathbf{n}} = \frac{1}{k_0} \nabla \Phi \tag{5.38}$$

where k_0 is constant, we must have

$$\nabla \times (\gamma \hat{\mathbf{n}}) = 0 \tag{5.39}$$

and therefore by Stokes' theorem (using the notation defined in Figure 5.3)

$$\int_{\Omega} \nabla \times (\gamma \hat{\mathbf{n}}) \cdot d\mathbf{a} = \int_{\delta\Omega} (\gamma \hat{\mathbf{n}}) \cdot d\mathbf{l} = 0 \tag{5.40}$$

if $\nabla \Phi$ is continuous and where the integral is around any closed path. This result is called "Lagrange's integral invariant."

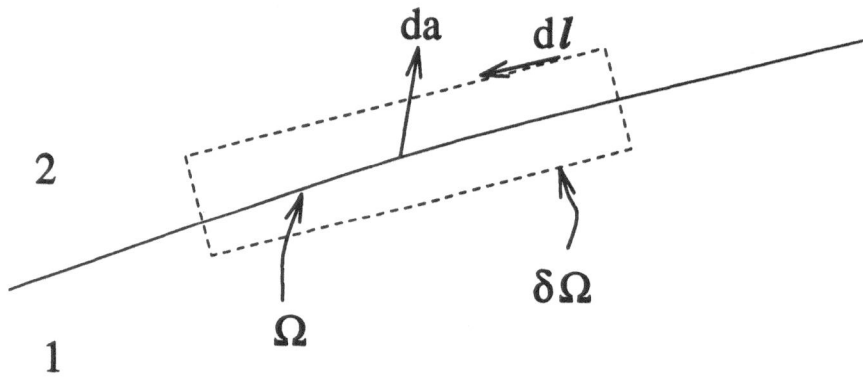

Fig. 5.3. Stokes pillbox around an interface between two media. Since the shape is arbitrary, we may choose the sides perpendicular to the interface to be infinitesimally small.

We can immediately use this to prove that the ray path connecting two points P_1 and P_2 is the minimum time path.[2] Figure 5.4 shows a bundle of rays one of which connects the two points P_1 and P_2. These rays are intersected at right angles by two surfaces of constant phase. To get the general result that the length of C is inevitably greater than or equal to the length of $\bar{C}$, and hence that the travel time $T(C)$ is greater than or equal to $T(\bar{C})$, it suffices to show that

$$(\gamma dl)_{\bar{Q}_1 \bar{Q}_2} \leq (\gamma dl)_{Q_1 Q_2}. \tag{5.41}$$

[2] This discussion comes straight out of Born and Wolf [8]. *Nota bene*, however the typo in their Equation (5) of Section 3.3.2.

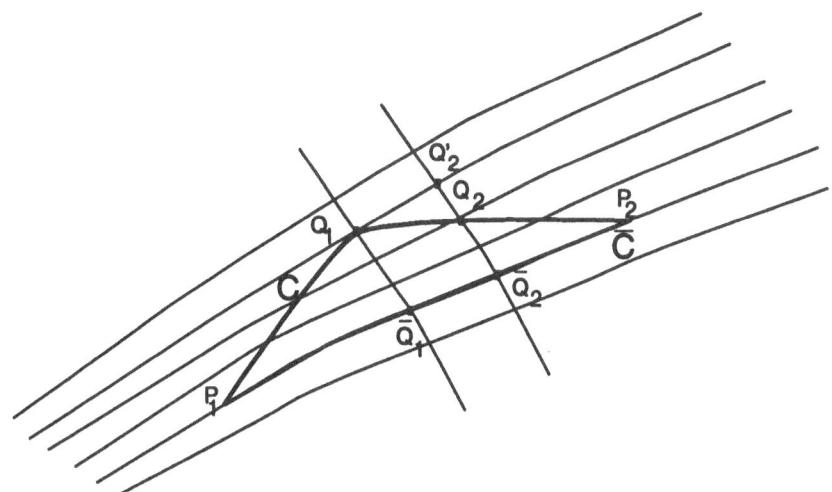

Fig. 5.4. The travel time along the ray path $\tilde{C}$ connecting two points P_1 and P_2 is less than for any other path such as C.

Using Lagrange's invariant on the infinitesimal triangle defined by $Q_1 Q_2 Q_2'$ we have

$$\int_{Q_1 Q_2 Q_2'} \gamma \hat{n} \cdot d\mathbf{l} = (\gamma \hat{n} \cdot d\mathbf{l})_{Q_1 Q_2} + (\gamma \hat{n} \cdot d\mathbf{l})_{Q_2 Q_2'} - (\gamma dl)_{Q_1 Q_2'} = 0. \qquad (5.42)$$

The last term doesn't involve any dot product because along the segment $Q_1 Q_2'$ the line element $d\mathbf{l}$ and the tangent to the ray (the normal of the isophasal surface) $\hat{n}$ are parallel. Similarly, along the path $Q_2 Q_2'$ the line element is perpendicular to the ray. Therefore $(\gamma \hat{n} \cdot d\mathbf{l})_{Q_2' Q_2} = 0$. Thus we have

$$(\gamma \hat{n} \cdot d\mathbf{l})_{Q_1 Q_2} = (\gamma dl)_{Q_1 Q_2'}. \qquad (5.43)$$

Now, for any path

$$(\gamma \hat{n} \cdot d\mathbf{l})_{Q_1 Q_2} \leq (\gamma dl)_{Q_1 Q_2}. \qquad (5.44)$$

Therefore

$$(\gamma dl)_{Q_1 Q_2} \geq (\gamma dl)_{Q_1 Q_2'} = (\gamma dl)_{\bar{Q}_1 \bar{Q}_2}. \qquad (5.45)$$

The last equality holds because the phase difference between two isophasal surfaces along any two raypaths must be the same. Thus we have shown that

$$\int_C \gamma dl \geq \int_{\tilde{C}} \gamma dl \qquad (5.46)$$

and, assuming constant speed, $T(C) \geq T(\bar{C})$, Fermat's principle.

Also note that although the equality strictly holds only when both paths are the same, the amount of inequality when C is not the same as $\bar{C}$ depends upon the amount by which

$(\gamma \hat{n}) \cdot (dl)_{Q_1 Q_2}$ differs from $\gamma |dl|_{Q_1 Q_2}$.

As you can easily show, when $\hat{n}$ and dl differ in directions by only a small angle β (that is, when C differs only slightly in direction from the local ray), then

$(\gamma \hat{n}) \cdot (dl)_{Q_1 Q_2} = (1 - \beta^2 - \cdots) \gamma |dl|_{Q_1 Q_2}$

so the inequality is only affected to second order in β. That is why the set of paths close to a ray path all have the same travel-time to first order in the deviation from the ray path.

5.3.1 Boundary Conditions and Snell's Laws

When we derived Lagrange's invariant the region Ω is completely arbitrary. However, let's consider the case in which this region surrounds a boundary separating two different media. If we let the sides of the Stokes pillbox perpendicular to the interface go to zero, then only the parts of the line integral tangential to the interface path contribute. And since they must sum to zero, that means that the tangential components must be continuous:

$$(\gamma_2 \hat{n}_2 - \gamma_1 \hat{n}_1)_{\text{tangential}} = 0 \qquad (5.47)$$

where the subscripts 1 and 2 refer to a particular side of the boundary. This is equivalent to saying that the vector

$$\gamma_1 \hat{n}_1 - \gamma_2 \hat{n}_2 \qquad (5.48)$$

must be normal to the boundary.

Now imagine a ray piercing the boundary and going through our Stokes pillbox as shown in Figure 5.5. If θ_1 and θ_2 are the angles of incidence and transmission, measured from the normal through the boundary, then continuity of the tangential components means that $\gamma_2 \sin \theta_2 = \gamma_1 \sin \theta_1$, which is called Snell's law of refraction. Similarly one can show that in the case of a reflected ray, the angle of incidence must equal the angle of reflection.

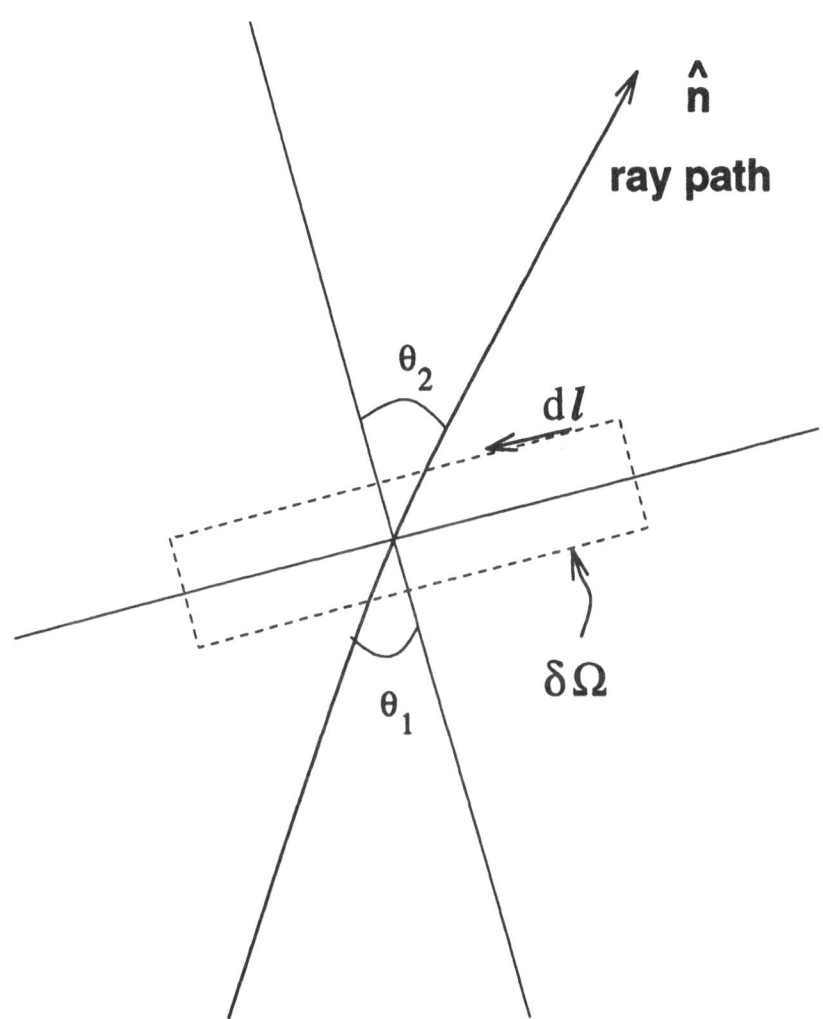

Fig. 5.5. Geometry for Snell's law.

Strictly speaking the above argument is not legal. If there are true discontinuities in the medium, then the conditions under which the ray approximation are valid (slow variation in material properties relative to the wavelength of the ray) are violated. To be precise we need to consider the boundary conditions of the full wave equation. It turns out that we will get the same result; so we can waffle a little bit and say something to the effect that what we are really talking about are plane waves refracting/reflecting at the boundary. To see that this is true, consider a plane pressure wave of unit amplitude ψ_i in 2D incident upon a vertical boundary at $x = 0$ at some angle of incidence θ_i

$$\psi_i = e^{ik_1(\cos\theta_i x + sin\theta_i z)} \tag{5.49}$$

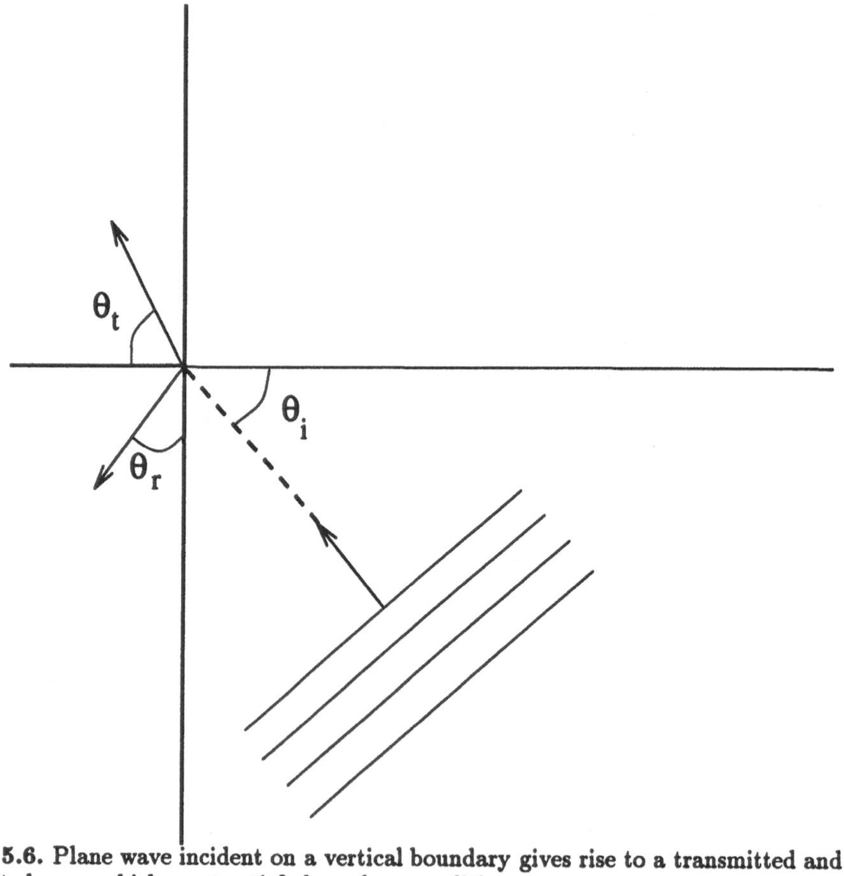

Fig. 5.6. Plane wave incident on a vertical boundary gives rise to a transmitted and a reflected wave which must satisfy boundary conditions.

Similarly for the reflected and transmitted plane waves we have (cf. Figure 5.6 for the geometry)

$$\psi_r = C_1 e^{ik_1(\cos\theta_r x + \sin\theta_r z)} \tag{5.50}$$

and

$$\psi_t = C_2 e^{ik_2(\cos\theta_t x + \sin\theta_t z)} \tag{5.51}$$

Convince yourself that ψ_i and ψ_r must have the same wavenumber and that all three plane waves must have the same frequency. Now, in order for the pressure to be continuous across the boundary,[3] we must have

$$[\psi_i + \psi_r = \psi_t]_{x=0} . \tag{5.52}$$

The only way that this can be true for arbitrary z is if: $k_1 \sin\theta_i = k_1 \sin\theta_r = k_2 \sin\theta_t$. In other words, $\theta_i = \theta_r$ and $\sin\theta_t = (c_2/c_1)\sin\theta_i$.

5.4 Fermat Redux

We end this section by showing that Fermat's principle of least time leads back precisely to the ray equations which we derived before by making a high-frequency asymptotic approximation. This discussion is patterned on the one in [5]. Let us begin by writing the travel time between any two points A and B along the ray as

$$t = \int_{\text{ray path}} \gamma(\mathbf{r}(\mu)) \, d\sigma(\mu) \tag{5.53}$$

where μ is a dimensionless parameter that increases monotonically along the ray. Using dots to denote differentiation with respect to path length we have

$$d\sigma = \sqrt{\dot{x}^2 + \dot{y}^2 + \dot{z}^2} \, d\mu = |\dot{\mathbf{r}}| \, d\mu. \tag{5.54}$$

In terms of μ the travel time can be written

$$t = \int_{\mu(A)}^{\mu(B)} \gamma(\mathbf{r}, \dot{\mathbf{r}}) |\dot{\mathbf{r}}| \, d\mu \tag{5.55}$$

$$= \int_{\mu(A)}^{\mu(B)} f(\mathbf{r}, \dot{\mathbf{r}}) \, d\mu \tag{5.56}$$

---

[3] See [1], Section 2.1 for discussion of this issue. The other relevant boundary condition is continuity of displacement for a welded contact, or merely continuity of normal displacement for the contact between an inviscid fluid and a solid.

where $f = \gamma |\dot{\mathbf{r}}|$. Fermat's principle, that the ray is (locally) the least time path, implies that variations in t must be zero:

$$\delta t = \int_{\mu(A)}^{\mu(B)} [\nabla_{\mathbf{r}} f \cdot \delta\mathbf{r} + \nabla_{\dot{\mathbf{r}}} f \cdot \delta\dot{\mathbf{r}}] \, d\mu = 0. \tag{5.57}$$

Integrating this by parts gives

$$\delta t = \int_{\mu(A)}^{\mu(B)} \left[\nabla_{\mathbf{r}} f - \frac{d}{d\mu} \nabla_{\dot{\mathbf{r}}} f \right] \cdot \delta\mathbf{r} \, d\mu = 0. \tag{5.58}$$

In order that this integral be zero for arbitrary variations $\delta\mathbf{r}$ it is necessary that the part of the integral within square brackets be identically zero:

$$\nabla_{\mathbf{r}} f - \frac{d}{d\mu} \nabla_{\dot{\mathbf{r}}} f = 0. \tag{5.59}$$

This is the Euler-Lagrange equaton of variational calculus [37].

Using the definition of the function f as $\gamma |\dot{\mathbf{r}}|$ it follows that

$$\nabla_{\mathbf{r}} f \;=\; |\dot{\mathbf{r}}| \nabla\gamma \tag{5.60}$$

$$\nabla_{\dot{\mathbf{r}}} f \;=\; \gamma \frac{\dot{\mathbf{r}}}{|\dot{\mathbf{r}}|}. \tag{5.61}$$

Finally we observe that $d\sigma = |\dot{\mathbf{r}}| \, d\mu$ so that the stationarity of t implies

$$\frac{d}{d\sigma} \left(\gamma \frac{d\mathbf{r}}{d\sigma} \right) = \nabla\gamma \tag{5.62}$$

which is none other than Equation 5.33. Thus we have shown that asymptotic ray theory and Fermat's principle lead to the same concept of rays.

Exercises

5.1 Using the ray equation

$$d_\sigma(\gamma d_\sigma \mathbf{r}) = \nabla\gamma$$

show that in a homogeneous medium, rays have the form of straight lines.

5.2 Show that in a medium in which γ depends only on depth z the quantity

$$\hat{z} \times \gamma \frac{d\mathbf{r}}{d\sigma}$$

is constant along rays. This shows that rays are confined to vertical planes and that the *ray parameter* $p = \gamma(z)\sin i(z)$ is constant, where $i(z)$ is the angle between the ray and the z axis.

5.3 Suppose that a medium has a spherically symmetric index of refraction (or slowness). Show that all rays are plane curves situated in a plane through the origin and that along each ray $\gamma r \sin \phi =$ constant, where ϕ is the angle between the position vector $\mathbf{r}$ and the tangent to the ray at the point $\mathbf{r}$. You may recognize this result as the analog of conservation of angular momentum of a particle moving in a central force.

5.4 What two independent equations must the amplitude P and the phase Φ satisfy, in order that the Helmholtz equation hold?

5.5 Next, show that if $\frac{\nabla^2 P}{P}$ can be neglected in comparison to $(\nabla\Phi)^2$, then one of the two equations you just derived reduces to $(\nabla\Phi)^2 = \frac{\omega^2}{c^2}$, the eikonal equation.

5.6 Use Lagrange's invariant to prove Snell's law of refraction.

6. Kirchhoff Migration

6.1 The Wave Equation in Disguise

In this first pass through the Kirchhoff migration algorithm we will restrict attention to the constant velocity case and follow, more or less, Schneider's original derivation [41]. In later chapters we will discuss generalization to variable velocity media. The derivation itself begins with the forward problem for the wave equation. Although this discussion will focus on the acoustic wave equation, it applies equally well to the migration of elastic or even electromagnetic scattering data.

Suppose that we wish to solve a boundary value problem for the scalar wave equation on the interior of some closed volume Ω in $\mathcal{R}^3$, with boundary conditions specified on a smooth boundary $\partial\Omega$. We're looking for a function ψ such that:

$$\nabla^2\psi - \frac{1}{c^2}\partial_t^2\psi = 0 \quad \forall \mathbf{r} \in \Omega. \tag{6.1}$$

The geometry is illustrated in Figure 6.1.

A complete specification of the problem requires that we specify the initial values of ψ and its normal derivative $\partial\psi/\partial n$ in Ω and either ψ or $\partial\psi/\partial n$ on $\partial\Omega$. Specifying the function on the boundary is called the Dirichlet problem; specifying the normal derivatives is called the Neumann problem. Some differential equations allow one to specify both types of conditions simultaneously (Cauchy problem), but the wave equation does not [37]. For starters, we will imagine our sources lying on the free surface of the Earth, which is going to be part of $\partial\Omega$, so that we can use the homogeneous (source-free) form of the wave equation.

Now when we use Huygen's principle, we are synthesizing a general solution of the wave equation from the cumulative effects of an infinite number of point sources. This can be made mathematically systematic by defining an impulse response function or "Green function"[1] which is the solution of the wave equation with a point source term. Irrespective of the boundary conditions involved, we define a

---

[1] We have decided to swim upsteam and abandon the awkward and singular Green's function in favor of Green function.

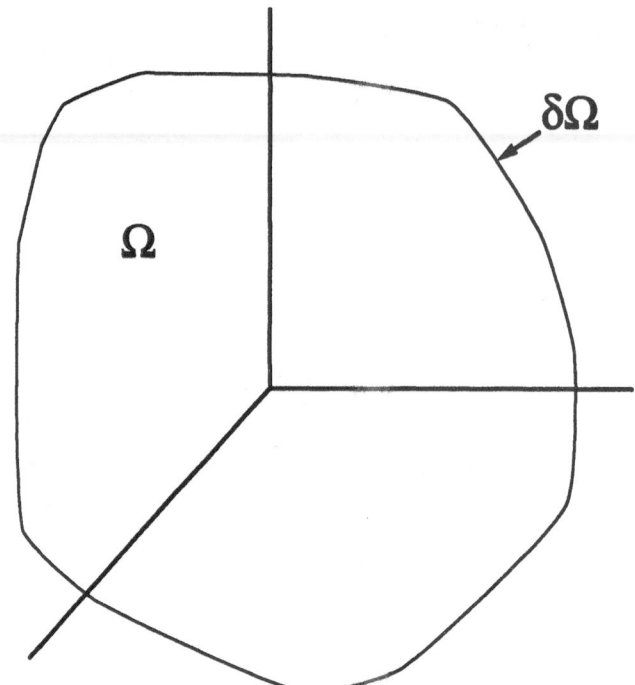

Fig. 6.1. Volume and surface defined in the specification of the wave equation.

Green function, depending on both source and observer coordinates, $\Gamma(\mathbf{r}, t; \mathbf{r}', t')$ as the solution of:

$$\nabla^2 \Gamma - \frac{1}{c^2}\partial_t^2 \Gamma = -4\pi\delta(\mathbf{r} - \mathbf{r}')\delta(t - t') \tag{6.2}$$

where δ is the Dirac delta function and $(\mathbf{r}', t')$ are the source coordinates.

If we multiply this equation by ψ and the wave equation by Γ and subtract one equation from the other we get

$$4\pi\delta(\mathbf{r} - \mathbf{r}')\delta(t - t')\psi = \Gamma\nabla^2\psi - \psi\nabla^2\Gamma + \frac{1}{c^2}\left(\Gamma\partial_t^2\psi - \psi\partial_t^2\Gamma\right) \tag{6.3}$$

or

$$4\pi\delta(\mathbf{r} - \mathbf{r}')\delta(t - t')\psi = \nabla \cdot [\Gamma\nabla\psi - \psi\nabla\Gamma] + \frac{1}{c^2}\partial_t\left[\Gamma\partial_t\psi - \psi\partial_t\Gamma\right] \tag{6.4}$$

To get rid of the delta function, we'd like to integrate this equality over some space-time volume. Ω is an obvious choice for the space part, and we'll integrate over all time:

$$\xi(\mathbf{r})\psi(\mathbf{r},t) \;=\; \int_{-\infty}^{\infty}\!\!\int_{\Omega} \nabla\cdot[\Gamma\nabla\psi - \psi\nabla\Gamma]\,dv'\,dt'$$

$$+\; \frac{1}{c^2}\int_{-\infty}^{\infty}\!\!\int_{\Omega} \partial_t\,[\Gamma\partial_t\psi - \psi\partial_t\Gamma]\,dv'\,dt' \quad \forall \mathbf{r}\in\mathcal{R}^3. \tag{6.5}$$

The left-hand side comes from integrating the delta function. The value of $\xi(\mathbf{r})$ is calculated by a limiting argument and turns out to be 4π if $\mathbf{r}$ is in the interior of Ω, 2π if $\mathbf{r}$ is on the boundary, and 0 otherwise.

The first integral on the right of Equation (6.5) can be converted to a surface integral over $\partial\Omega$ via the divergence theorem. The second integral can be integrated by parts with respect to time giving a term of the form

$$\left[\int_{\Omega}\Gamma\partial_t\psi - \psi\partial_t\Gamma\,dv'\right]_{-\infty}^{+\infty}. \tag{6.6}$$

Clearly we can assume that ψ and $\partial_t\psi$ are zero until sometime after the source is fired off, say at $t'=0$, so the lower limit is zero. And provided the Green function is causal or assuming a Sommerfeld radiation condition on ψ, the upper limit must be zero too.[2] The result of all this is the Kirchhoff Integral Theorem:

$$\xi(\mathbf{r})\psi(\mathbf{r},t) = \int_{0}^{\infty}\!\!\int_{\partial\Omega} [\Gamma\nabla\psi - \psi\nabla\Gamma]\cdot\mathbf{n}\,da'\,dt' \quad \forall \mathbf{r}\in\mathcal{R}^3 \tag{6.7}$$

where $\mathbf{n}$ is the (outward pointing) unit normal of $\partial\Omega$.

The Kirchhoff Integral Theorem gives the solution of the wave equation ψ everywhere in space once the values of ψ and its normal derivative are known on the boundary. But since we cannot in general specify both of these consistently (Cauchy conditions), we need to solve for one in terms of the other. This we do by taking the limit of Equation (6.7) as the observation point $\mathbf{r}$ approaches the surface. Using the notation:

$$\psi|_{\partial\Omega} \;=\; f \tag{6.8}$$

$$\frac{\partial\psi}{\partial n}\Big|_{\partial\Omega} \;=\; g \tag{6.9}$$

we then have for the Neumann problem

$$f(\mathbf{r},t) = \frac{1}{2\pi}\int_{0}^{\infty}\!\!\int_{\partial\Omega}[\Gamma g - f\partial_n\Gamma]\,da'\,dt' \quad \forall \mathbf{r}\in\partial\Omega \tag{6.10}$$

So to recap, since we can't specify both the function and its normal derivative, we must specify one and solve for the other. If we specify, for example, the normal

---

[2] This will come back to haunt us later.

derivative, i..e, g, then Equation (6.10) is an *integral equation* for f. If we then solve for f we can use the Kirchhoff Integral to compute the solution to the wave equation. There is a comparable equation such as Equation (6.10) for the Dirichlet problem, where we specify f and solve for g.

So far we haven't said anything about what boundary conditions the Green function must satisfy. But if you look back over the derivations thus far, we haven't assumed anything at all in this regard. The expressions derived so far are true for any Green function whatsoever, i.e., any function satisfying Equation (6.2). Now in special cases it may be possible to find a Green function which vanishes on part or all of the boundary in question, or whose normal derivative does. If the Green function vanishes on $\delta\Omega$ then we can solve the Dirichlet problem for the wave field directly from Kirchhoff's Integral since the term in that integral involving the unknown normal derivative of the wave field is cancelled. In other words, if the Green function vanishes on the boundary then Equation (6.7) reduces to:

$$\psi(\mathbf{r},t) = \frac{1}{4\pi} \int_0^\infty \int_{\partial\Omega} f(\mathbf{r}',t')\partial_n \Gamma(\mathbf{r},t;\mathbf{r}',t') \, da' \, dt' \tag{6.11}$$

which is not an integral equation at all since by assumption we know f. Similarly, if the normal derivative of the Green function vanishes on the boundary then we can immediately solve the Neumann problem for the wave field since the unknown term involving its boundary values is cancelled. In either event, having a Green function satisfying appropriate boundary conditions eliminates the need to solve an integral equation. If we don't have such a special Green function, we're stuck with having to solve the integral equation somehow, either numerically or by using an approximate Green function as we will do later when we attempt to generalize these results to nonconstant wavespeed media.

6.2 Application to Migration

Consider now the problem of data recorded on the surface $z = 0$. To begin, we will consider the case of media with constant wavespeed, so that there are no boundaries in the problem. Of course, this is really a contradiction; we'll see how to get around this difficulty later. Then we can take $\partial\Omega$ to consist of the plane $z = 0$ smoothly joined to an arbitrarily large hemisphere extending into the lower half-space. If we let the radius of the hemisphere go to infinity, then provided the contribution to the integral from this boundary goes to zero sufficiently rapidly, the total integral over $\partial\Omega$ reduces to an integral over the plane $z = 0$. Showing that the integral over the lower boundary can really be neglected is actually quite tricky since even if we use a Green function which satisfies Sommerfeld, the field itself may not. We will discuss this in detail later in Paul Docherty's talk on Kirchhoff inversion.

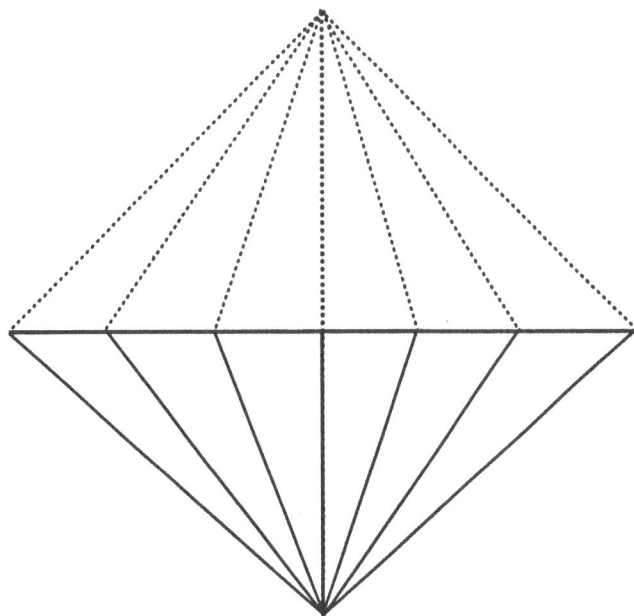

Fig. 6.2. A Green function which vanishes on a planar surface can be calculated by the method of images.

In the special case of a constant velocity Earth, two Green functions which vanish on the surface $z = 0$ can be calculated by the method of images. They are:

$$\Gamma_r(\mathbf{r},t;\mathbf{r}',t') = \frac{\delta(t - t' - R/c)}{R} - \frac{\delta(t - t' - R'/c)}{R'} \tag{6.12}$$

and

$$\Gamma_a(\mathbf{r},t;\mathbf{r}',t') = \frac{\delta(t - t' + R/c)}{R} - \frac{\delta(t - t' + R'/c)}{R'} \tag{6.13}$$

where

$$R = \sqrt{(x - x')^2 + (y - y')^2 + (z - z')^2}$$

and

$$R' = \sqrt{(x - x')^2 + (y - y')^2 + (z + z')^2}.$$

We won't derive these Green functions here; that will be left for an exercise. However it is obvious from our discussion of spherical waves that these are spherically symmetric solutions of the wave equation. The subscripts a and r refer, respectively, to advanced and retarded. The retarded Green function is causal, propagating outward from its origin as t increases. The advanced Green function is anti-causal. It represents a converging spherical wave propagating backwards in time.

Now all we have to do is plug one of these Green functions into Equation (6.11). And since migration involves propagating diffraction hyperbolae back to their origin, it is clear that we should be using Γ_a. Notice too that $R'(z') = R(-z')$, so that we can combine the normal (i.e., z) derivatives of the two delta functions with the result (for $\mathbf{r}$ in the subsurface):

$$\psi(\mathbf{r},t) = \frac{-1}{2\pi} \int_0^\infty \int_{z'=0} f(\mathbf{r}',t') \partial_{z'} \frac{\delta(t-t'+R/c)}{R} \, da' \, dt'. \qquad (6.14)$$

And since z appears only in the combination $z - z'$

$$\psi(\mathbf{r},t) = \frac{1}{2\pi} \partial_z \int_0^\infty \int_{z'=0} f(\mathbf{r}',t') \frac{\delta(t-t'+R/c)}{R} \, da' \, dt'. \qquad (6.15)$$

Finally, using the properties of the delta function to collapse the time integration, we have:

$$\psi(\mathbf{r},t) = \frac{1}{2\pi} \partial_z \int_{z'=0} \frac{f(\mathbf{r}',t+R/c)}{R} \, da'. \qquad (6.16)$$

This is Schneider's Kirchhoff migration formula [41].

This has been a rather involved derivation to arrive at so simple looking a formula, so let's recap what we've done. We started by converting the wave equation into an integral equation by introducing an arbitrary Green function. We observed that if we were clever, or lucky, enough to find a particular Green function which vanished (or whose normal derivative did) on the surface that we were interested in, then the integral equation collapsed to an integral relation which simply mapped the (presumed known) boundary values of the wave field into a solution valid at all points in space. For the problem of interest to us, data recorded at $z = 0$, it turned out that we were able to write down two such special Green functions, at least assuming a constant velocity Earth. The question now is what boundary values to we use?

We have already observed that CMP stacking produces a function $\psi_s(x,y,z=0,t)$ whose value at each mid-point location approximates the result of a single, independent zero-offset seismic experiment. So, the boundary data for the migration procedure do not correspond to the observations of a real physical experiment and the solution to the initial-boundary value problem that we have worked so hard on *is not an observable wave field.* From an abstract point of view we may simply regard the Kirchhoff migration formula, Equation (6.16), as a mapping from the space of possible boundary values into solutions of the wave equation.

For post-stack migration, we identify f in Equation (6.16) with the CMP stacked data ψ_s. Then we set $c \to c/2$ and $t = 0$ for the exploding reflector model. And finally we perform the integration over the recording surface x', y':

$$\text{depth section} \equiv \psi(\mathbf{r}) = \frac{1}{2\pi} \partial_z \int_{z'=0} \frac{\psi_s(\mathbf{r}', 2R/c)}{R} \, da'. \tag{6.17}$$

The normal derivative outside of the integral is responsible for what we called the "obliquity factor" in our discussion of the Huygens-Fresnel principle. Remember, R is the distance between the output point on the depth section and the particular receiver or trace location on the surface $z' = 0$. So as we integrate along the surface $z' = 0$ we are really summing along diffraction hyperbolae in the zero-offset data. This is illustrated in Figure 6.3.

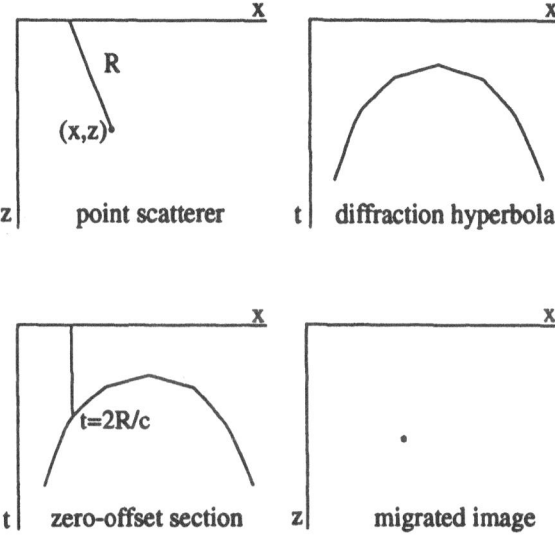

Fig. 6.3. A point scatterer in the Earth gives rise to a diffraction hyperbola on the time section. Kirchhoff migration amounts to a weighted sum along this hyperbola with the result going into a particular location on the depth section.

Equation (6.17) can also be viewed as a spatial convolution which downward propagates the recorded data from one z level to the next. Mathematically there is nothing special about $z = 0$; it just happens to be where we recorded the data. So if we were faced with a layered medium, one in which the velocity were constant within each layer, we could use this integral to downward propagate the recorded data one layer at a time. Since the Fourier transform of a convolution is just the product of the Fourier transforms, it makes sense to re-cast this downward propagation operation in the spatial Fourier or wavenumber domain. In the appendix to his paper [41] Schneider shows that the Fourier transform of Equation (6.17) can be written:

$$\psi(k_x, k_y, z + \Delta z, \omega) = \psi(k_x, k_y, z, \omega) H(k_x, k_y, \Delta z, \omega) \tag{6.18}$$

where

$$H = e^{\pm i \Delta z \sqrt{\left(\frac{\omega}{c}\right)^2 - k_x^2 - k_y^2}}. \tag{6.19}$$

This means that for constant velocity media, downward propagation is purely a phase shift operation.

Another commonly seen form of the Kirchhoff migration procedure is obtained by performing the z differentiation. Using the chain rule we find that:

$$\psi(\mathbf{r}) = \frac{1}{2\pi} \int_{z'=0} \frac{\cos\theta}{Rc} \left[\partial_{t'} \psi_s + \frac{c}{R} \psi_s \right]_{t'=2R/c} da' \tag{6.20}$$

where θ is the angle between the z axis and the line joining the ouput point on the depth section (x, y, z) and the receiver location $(x', y', 0)$. Because of the R in the denominator, you will often see the second term inside the brackets ignored.

All of these formulae for Kirchhoff migration can be interpreted as summing along diffraction curves. We look upon every point in the depth section, say $\mathbf{r}$, as a possible point diffractor. Since we, presumably, know the velocity above this point, we can find the apex of its hypothetical diffraction hyperbola. Then we sum along this curve according to an integral such as Equation (6.20) and deposit the value so obtained at the point $\mathbf{r}$. But Kirchhoff migration is significantly more accurate than simply summing along diffraction hyperbolae, even in constant velocity media, because of the presence of the obliquity factor $(\cos\theta)$ and the derivative of the data.

When the wavespeed c is not constant, the Kirchhoff formulae must be generalized. Essentially this amounts to replacing the straight ray travel time R/c with a travel time accurately computed in the particular medium:

$$t = T(x, y, z; x', y', z' = 0); \tag{6.21}$$

or, what amounts to the same thing, using a Green function valid for a variable velocity medium. We will discuss this later in the section on Kirchhoff-type inversion formulae.

6.3 Summary of Green Functions

The free-space Green functions for the Helmholtz equation

$$\nabla^2 G_k(\mathbf{r}, \mathbf{r}') + k^2 G_k(\mathbf{r}, \mathbf{r}') = -4\pi\delta(\mathbf{r} - \mathbf{r}') \tag{6.22}$$

are:

$$
\begin{aligned}
G_k(\mathbf{r}, \mathbf{r}') &= \frac{e^{ik|\mathbf{r}-\mathbf{r}'|}}{|\mathbf{r} - \mathbf{r}'|} &\tag{6.23}\\
&= i\pi H_0^{(1)}(k|\mathbf{r} - \mathbf{r}'|) &\tag{6.24}\\
&= \frac{2\pi i}{k} e^{ik|x-x'|} &\tag{6.25}
\end{aligned}
$$

in, respectively 3, 2, and 1 dimensions, where $H_0^{(1)}$ is the zero-th order Hankel function of the first kind.

The free-space Green functions for the wave equaton

$$\nabla^2 G(\mathbf{r}, t; \mathbf{r}', t') - \frac{1}{c^2}\partial_t^2 G(\mathbf{r}, t; \mathbf{r}', t) = -4\pi\delta(\mathbf{r} - \mathbf{r}')\delta(t - t') \tag{6.26}$$

are:

$$
\begin{aligned}
G(\mathbf{r}, t; \mathbf{r}', t') &= \frac{1}{|\mathbf{r} - \mathbf{r}'|}\delta\left[|\mathbf{r} - \mathbf{r}'|/c - (t - t')\right] &\tag{6.27}\\
&= \frac{2c}{\sqrt{c^2(t-t') - |\mathbf{r} - \mathbf{r}'|^2}}\Theta\left((t - t') - \frac{|\mathbf{r} - \mathbf{r}'|}{c}\right) &\tag{6.28}\\
&= 2c\pi\Theta\left((t - t') - \frac{|x - x'|}{c}\right) &\tag{6.29}
\end{aligned}
$$

in, respectively 3, 2, and 1 dimensions, where Θ is the step function: $\Theta(x) = 0$ for $x < 0$ and $\Theta(x) = 1$ for $x > 0$. For more details see Chapter 7 of Morse and Feshbach [37].

Exercises

6.1 In this extended exercise you will derive, following [30], the free-space Green function for both the Helmholtz equation and the wave equation. First we define $G(\mathbf{r}, \mathbf{r}', k)$ to be a solution to the Helmholtz equation with a point source:

$$\left(\nabla^2 + k^2\right) G(\mathbf{r}, \mathbf{r}', k) = -4\pi\delta(\mathbf{r} - \mathbf{r}').$$

Now show that if there are no boundaries involved, G must depend only on $r - r'$ and must in fact be spherically symmetric. Define $\mathbf{R} = \mathbf{r} - \mathbf{r}'$ and $R = |\mathbf{R}|$. Next, show that the Helmholtz equation reduces to

$$\frac{1}{R}\frac{d^2}{dR^2}(RG) + k^2 G = -4\pi\delta(\mathbf{R}).$$

Show that everywhere but $R = 0$ the solution of this equation is

$$RG(R) = Ae^{ikR} + Be^{-ikR}.$$

The delta function only has influence around $R = 0$. But as $R \to 0$ the Helmholtz equation reduces to Poisson's equation, which describes the potential due to point charges or masses. From this you should deduce the constraint: $A + B = 1$. The particular choice of A and B depend on the time boundary conditions that specify a particular problem. In any event, you have now shown that the free-space Green function for the Helmholtz equation is

$$G(R) = A\frac{e^{ikR}}{R} + B\frac{e^{-ikR}}{R}$$

with $A + B = 1$. It is convenient to consider the two terms separately, so we define

$$G^{\pm}(R) = \frac{e^{\pm ikR}}{R}.$$

Now show that the time-domain Green function obtained from the Fourier transform

$$G^{\pm}(R, t - t') \equiv G^{\pm}(R, \tau) = \frac{1}{2\pi}\int_{-\infty}^{\infty}\frac{e^{\pm ikR}}{R}e^{-i\omega\tau}\,d\omega$$

where $\tau = t - t'$ is the relative time between the source and observation point. Now what is the final form in the time domain of the Green functions $G^{\pm}(\mathbf{r}, t; \mathbf{r}', t') = G^{\pm}(\mathbf{r} - \mathbf{r}'; t - t')$? Which one of these is causal and which one is anti-causal and why?

For more details, the interested reader is urged to consult the chapter on Green function methods in Volume I of Morse and Feshbach [37].

6.2 Go through the same derivation that leads up to Schneider's constant velocity migration formula, but use the 2D Green function given above. You should end up with a migrated depth section that looks like:

$$\psi(x,z) = -\frac{1}{\pi}\partial_z \int dx' \int_{R/c} \frac{\psi_s(x,0,t')}{\sqrt{t'^2 - R^2/c^2}}\, dt'.$$

This result is often written in the following far-field form

$$\psi(x,z) = \int \frac{\cos\theta}{2\pi Rc}\partial_t^{1/2}\psi_s(x,0,R/c)\, dx'$$

where $\partial_t^{1/2}$ is a fractional derivative. The half-derivative operator is most easily implemented in the frequency domain where it amounts to multiplication by $\sqrt{(i\omega)}$.

Fractional derivatives (and integrals) are defined in a purely formal fashion using Cauchy's theorem for analytic functions. Let f be analytic everywhere within and on a simple closed contour C taken in a positive sense. If s is any point interior to C, then

$$f(s) = \frac{1}{2\pi i}\int_C \frac{f(z)}{z-s}\, dz.$$

By differentiating this expression n times we get

$$f^{(n)}(s) = \frac{n!}{2\pi i}\int_C \frac{f(z)}{(z-s)^{n+1}}\, dz.$$

Now the expression on the right is well-defined for non-integral values of n provided we replace the factorial by the Gamma function.[3] Therefore we identify, again purely formally, the expression on the left as the fractional derivative (or fractional integral when n is negative) of the indicated order.

6.3 A point-source in 2D is equivalent to a line-source in 3D. Therefore in an infinite, homogeneous medium, the time-domain 2D Green function can be obtained by integrating the 3D Green function along a line as follows:

$$g(\mathbf{p},t;\mathbf{p}',t') = \int_{-\infty}^{\infty} \frac{\delta\left[(R/c)-(t-t')\right]}{R}\, dz',$$

---

[3] The Gamma function $\Gamma(z)$ is defined via the integral:

$$\Gamma(z) = \int_0^\infty t^{z-1}e^{-t}\, dt.$$

It can be shown that $\Gamma(1) = 1$ and $\Gamma(z+1) = z\Gamma(z)$. The Gamma function is analytic everywhere except at the points $z = 0, -1, -2, \ldots$ where it has simple poles. For more details, see [37].

where p and p' are radius vectors in the x, y plane, and

$$R = \sqrt{(x - x')^2 + (y - y')^2 + (z - z')^2}.$$

You should expect that g is a function only of $P = |p - p'|$ and $\tau = t - t'$.

Show that $g(P, \tau) = 2c/\sqrt{c^2\tau^2 - P^2}$ if $P < c\tau$ and $g(P, \tau) = 0$ otherwise. Hint: introduce the change of variables $\xi = z' - z$. Then

$$R^2 = \xi^2 + P^2; \qquad d\xi/dR = R/\xi.$$

These will allow you to do the integration over the R coordinate so that the delta function can be evaluated.

6.4 In cylindrical coordinates (ρ, ϕ, z) the Laplacian is

$$\nabla^2 f = \frac{1}{\rho} \frac{\partial}{\partial \rho} \left(\rho \frac{\partial f}{\partial \rho} \right) + \frac{1}{\rho^2} \frac{\partial^2 f}{\partial \phi^2} + \frac{\partial^2 f}{\partial z^2}.$$

Assuming an infinite (i.e., no boundaries), constant-velocity medium, Show that away from the source and at sufficiently low frequency, the frequency-domain (i.e., Helmholtz equation) Green function is proportional to $\log \rho$.

6.5 Schneider's Kirchhoff migration formula is

$$\text{depth section} = \frac{1}{2\pi} \partial_z \int_{z'=0} \frac{\psi_s(\mathbf{r}', t' = 2R/c)}{R} \, da'$$

where $R = \sqrt{(x - x')^2 + (y - y')^2 + (z - z')^2}$, ψ_s is the stacked data and the integral is over the recording surface. Show that in the far field (i.e., when R is large) the depth section is proportional to

$$\int_{z'=0} \frac{\cos \theta}{R} [\partial_{t'} \psi_s]_{t'=2R/c} \, da'.$$

7. Kirchhoff Migration/Inversion[1]

We will begin by looking at Claerbout's heuristic imaging condition, and show that this is equivalent to the formulae derived in the theory of Kirchhoff inversion. Inversion means that we provide a geometrical image of the reflectors, as well as quantitative estimates of the reflection coefficients. A source is excited at some

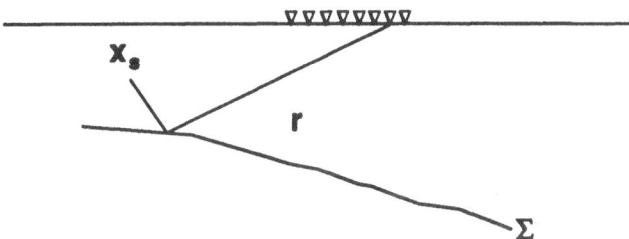

Fig. 7.1. The incident field reflects off a boundary Σ at some depth. The generic observation point is **r**.

point $\mathbf{x}_s$. Let u_I denote the incident field, and u_S the scattered field. Suppose we know the scattered field at some depth point **r**, as illustrated in Figure 7.1. Claerbout's imaging condition is that the migrated image at **r** is given by

$$m(\mathbf{r}) = \frac{1}{2\pi} \int d\omega F(\omega) \frac{u_S(\mathbf{r}, \mathbf{x}_S, \omega)}{u_I(\mathbf{r}, \mathbf{x}_S, \omega)} \tag{7.1}$$

where F is a filter emphasising the band-limited nature of the scattering field.

This looks as if we are deconvolving the scattered field by the incident field. If the medium is slowly varying (in the WKBJ sense), then we can approximate

$$u_I = A_I(\mathbf{r}, \mathbf{x}_s) e^{i\omega \tau_I(\mathbf{r}, \mathbf{x}_s)} \tag{7.2}$$

---

[1] This chapter is based on talks that Paul Docherty gave before the class in 1993 and 1994, which were adapted from his paper [18].

where τ_I is the travel time from the source to the observation point $\mathbf{r}$. In this case we have

$$m(\mathbf{r}) = \frac{1}{2\pi A_I} \int d\omega F(\omega) u_S(\mathbf{r}, \mathbf{x}_S, \omega) e^{-i\omega \tau_I(\mathbf{r}, \mathbf{x}_s)}. \tag{7.3}$$

This has the form of a Fourier transform and amounts to evaluating the scattered field at a time τ_I. If we continue along these lines and use the WKBJ approximation for the scattered field, then we have

$$m(\mathbf{r}) = \frac{A_S}{2\pi A_I} \int d\omega F(\omega) e^{-i\omega(\tau_I(\mathbf{r}, \mathbf{x}_s) - \tau_S(\mathbf{r}, \mathbf{x}_s))}. \tag{7.4}$$

But this integral is just a delta function, band-limited by the presence of F. Denoting this by δ_B, and identifying the ratio $\frac{A_S}{A_I}$ as the reflection coefficient, we have the fundamental result

$$m(\mathbf{r}) = R(\mathbf{r}) \delta_B(\tau_I - \tau_S). \tag{7.5}$$

The interpretation of this expression is clear: it is non-zero only where the reflection coefficient is non-zero, and then only when $\tau_I \approx \tau_S$. The last $\approx$ is because of the band-limited nature of the delta-function. The peak value certainly occurs when $\tau_I = \tau_S$ and is given by

$$m_{\text{peak}}(\mathbf{r}) = \frac{1}{2\pi} R(\mathbf{r}) \int d\omega F(\omega) \tag{7.6}$$

or

$$m_{\text{peak}}(\mathbf{r}) = \frac{1}{2\pi} R(\mathbf{r}) \times \text{Area of filter F.} \tag{7.7}$$

Now, we can pretend we know A_I and τ_I, but how do we get the scattered field at depth? The answer is: using Green functions. We will proceed in much the same way as in the preceeding chapter, except that now we take explicit account of the fact that the velocity is not constant. We will denote the true wavespeed by $v(\mathbf{x})$; it will be assumed that v is known above the reflector. We will introduce a "background" wavespeed $c(\mathbf{x})$ which is also nonconstant, but assumed to be known and equal to v above the reflector. By defintion, the total field u satisfies the wave equation with v:

$$\left[\nabla^2 + \frac{\omega^2}{v^2} \right] u(\mathbf{x}, \mathbf{x}_s) = -\delta(\mathbf{x} - \mathbf{x}_s) \tag{7.8}$$

whereas the "incident" field satisfies the wave equation with the background wavespeed

$$\left[\nabla^2 + \frac{\omega^2}{c^2}\right] u_I(\mathbf{x}, \mathbf{x}_s) = -\delta(\mathbf{x} - \mathbf{x}_s). \tag{7.9}$$

For this hypothetical experiment we can make the source whatever we want, so we'll make it a delta-function. This is without loss of generality since we can synthesize an arbitrary source by superposition.

The basic decomposition of the total field is:

$$u(\mathbf{x}, \mathbf{x}_s) = u_I(\mathbf{x}, \mathbf{x}_s) + u_S(\mathbf{x}, \mathbf{x}_s). \tag{7.10}$$

When we talked about the plane-wave boundary conditions in the chapter on rays, we decomposed the field into an incident and scattered (or reflected) field on one side of the boundary, and a transmitted field on the other side of the boundary. Here we are using a different decomposition; neither more nor less valid–just different. Below the reflector, the scattered field is defined to be the total field minus the incident field. We can imagine the incident field being defined below the reflector, just by continuing the background model c below the reflector. This is a mathematically well-defined concept, it just doesn't correspond to the physical experiment. As we will see though, it let's us get to the heart of the migration problem.

Introduce a quantity α which represents, in some sense, the perturbation between the background model and the true model:

$$\alpha \equiv \frac{c(\mathbf{x})^2}{v(\mathbf{x})^2} - 1 \tag{7.11}$$

so that

$$\frac{\alpha}{c^2} \equiv \frac{1}{v^2} - \frac{1}{c^2}. \tag{7.12}$$

Since $c = v$ above the reflector, α is zero in this region. Whereas α is nonzero below the reflector. In terms of α, the equation for the scattered field can be written:

$$\left[\nabla^2 + \frac{\omega^2}{c^2}\right] u_S(\mathbf{x}, \mathbf{x}_S, \omega) = -\frac{\omega^2}{c^2} \alpha u(\mathbf{x}, \mathbf{x}_S, \omega). \tag{7.13}$$

Now let us introduce a Green function for the background model:

$$\left[\nabla^2 + \frac{\omega^2}{c^2}\right] G(\mathbf{x},\mathbf{r},\omega) = -\delta(\mathbf{x} - \mathbf{r}) \tag{7.14}$$

As usual, we convert this into an integral relation via Green's identity

$$-u_S(\mathbf{r},\mathbf{x}_S,\omega) + \int_V G(\mathbf{x},\mathbf{r},\omega)\frac{\omega^2}{c^2}\alpha u(\mathbf{x},\mathbf{x}_S,\omega)dV$$

$$= \int_S \left[u_S(\mathbf{x},\mathbf{x}_S,\omega)\frac{\partial G(\mathbf{x},\mathbf{r},\omega)}{\partial n} - G(\mathbf{x},\mathbf{r},\omega)\frac{\partial u_S(\mathbf{x},\mathbf{x}_S,\omega)}{\partial n}\right] dS \tag{7.15}$$

At this point we have complete flexibility as to how we choose the volume/surface of integration. We also have yet to decide which Green function to use. For example, it could be causal or anti-causal. We have a great variety of legitimate choices, all giving rise to different formulae. Some of these formulae will be useful for migration and some will not. We have to see, essentially by trial and error which ones are which.

7.1 Case I: Causal G, $S = S_0 + S_{\infty+}$

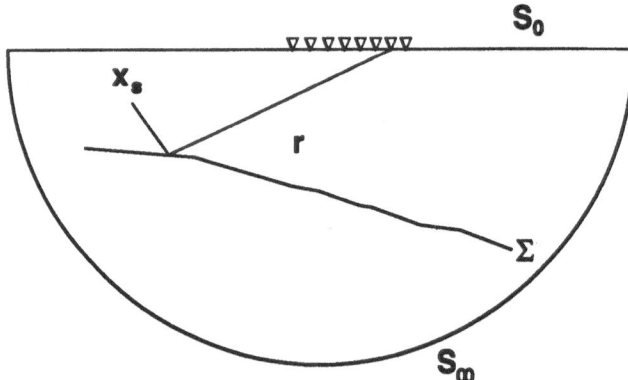

Fig. 7.2. Define the surface S to be the union of the recording surface S_0 and a surface joining smoothly with S_0 but receding to infinity in the $+z$ direction.

Let S be the union of S_0, the data recording surface ($z = 0$) and a surface $S_{\infty+}$ which is smoothly joined to S_0 but which recedes to infinity in the $+z$ direction (Figure 7.2).

First we observe that the surface integral over S_0 must be zero. To see this, apply Green's identity to the volume enclosed by $S = S_0 + S_{\infty-}$, the corresponding surface going to infinity in the upper half plane. Then in the volume enclosed by S, α is zero, and the scattered field is zero too since the Green function has no in support this domain. Therefore the total surface integral

$$\int_{S_0+S_{\infty-}} \left[u_S(\mathbf{x},\mathbf{x}_S,\omega) \frac{\partial G(\mathbf{x},\mathbf{r},\omega)}{\partial n} - G(\mathbf{x},\mathbf{r},\omega) \frac{\partial u_S(\mathbf{x},\mathbf{x}_S,\omega)}{\partial n} \right] dS \qquad (7.16)$$

is equal to zero.

But as $S_{\infty-}$ goes to infinity the fields G and u must go to zero by virtue of the Sommerfeld radiation condition (since in this case both G and u are causal). Therefore the surface integral over S_0 must be zero itself. So now if we use $S = S_0 + S_{\infty+}$, since the surface integral over $S_{\infty+}$ must be zero by the same Sommerfeld argument, we have that

$$u_S(\mathbf{r},\mathbf{x}_S,\omega) = \int_V G(\mathbf{x},\mathbf{r},\omega) \frac{\omega^2}{c^2} \alpha u(\mathbf{x},\mathbf{x}_S,\omega) dV. \qquad (7.17)$$

This expression requires that we know the total field $u(\mathbf{x},\mathbf{x}_S,\omega)$ and α everywhere within the volume V. If we approximate the total field by the incident field (the so-called Born approximation) then we have

$$u_S(\mathbf{r},\mathbf{x}_S,\omega) \approx \omega^2 \int_V G(\mathbf{x},\mathbf{r},\omega) \frac{\alpha}{c^2} u_I(\mathbf{x},\mathbf{x}_S,\omega) dV. \qquad (7.18)$$

which is a just a linear function of the unknown perturbation α. In other words we have $u_s \approx L(\alpha)$ where L is a linear function. So by choosing a causal Green function and a surface extending to infinity in the $+z$ direction, we end up with a Born inversion formula for the unknown wavespeed perturbation α.

7.2 Case II: Causal G, $S = S_0 + \Sigma$

If we now take the volume V to be the area bounded by recording surface and the reflecting surface, then the volume integral goes away since α is zero in this region. This gives

$$-u_S(\mathbf{r},\mathbf{x}_S,\omega) \qquad (7.19)$$

$$= \int_{S_0+\Sigma} \left[u_S(\mathbf{r},\mathbf{x}_S,\omega) \frac{\partial G(\mathbf{x},\mathbf{r},\omega)}{\partial n} - G(\mathbf{x},\mathbf{r},\omega) \frac{\partial u_S(\mathbf{x},\mathbf{x}_S,\omega)}{\partial n} \right] dS.$$

Now the integral over S_0 is zero by the same argument we used in Case I. So in fact,

$$-u_S(\mathbf{r}, \mathbf{x}_S, \omega) \tag{7.20}$$

$$= \int_\Sigma \left[u_S(\mathbf{x}, \mathbf{x}_S, \omega) \frac{\partial G(\mathbf{x}, \mathbf{r}, \omega)}{\partial n} - G(\mathbf{x}, \mathbf{r}, \omega) \frac{\partial u_S(\mathbf{x}, \mathbf{x}_S, \omega)}{\partial n} \right] dS.$$

But to make use of it we need the scattered field at depth. So we plug in the WKB expressions for $u_S(\mathbf{x}, \mathbf{x}_S, \omega) = A_S(\mathbf{x}, \mathbf{x}_S) e^{i\omega \tau_S(\mathbf{x}, \mathbf{x}_S)}$ and $G = A_G(\mathbf{r}, \mathbf{x}_s) e^{i\omega \tau_G(\mathbf{r}, \mathbf{x}_s)}$, then computing the normal derivatives is straightforward:

$$\partial_n G = i\omega \hat{\mathbf{n}} \cdot \nabla \tau_G A_G e^{i\omega \tau_G} \tag{7.21}$$

neglecting the gradient of the amplitude (WKBJ again) and

$$\partial_n u_S = i\omega \hat{\mathbf{n}} \cdot \nabla \tau_S A_S e^{i\omega \tau_S}. \tag{7.22}$$

Plugging these into the surface integral, we arrive at

$$-u_S(\mathbf{r}, \mathbf{x}_S, \omega) = \int_\Sigma i\omega A_S A_G \left[\hat{\mathbf{n}} \cdot \nabla \tau_G - \hat{\mathbf{n}} \cdot \nabla \tau_S \right] e^{i\omega[\tau_G + \tau_S]}. \tag{7.23}$$

This gives us a formula for doing WKBJ modeling: we take $A_S = RA_I$, where R is the reflection coefficient, and we would have to compute A_I, τ_G and $\tau_S(= \tau_I)$ on Σ via raytracing.

7.3 Case III: G Anticausal, $S = S_0 + \Sigma$

In this case the volume V is still the region between the recording surface S_0 and the reflecting surface Σ, so the volume integral goes to zero. What we would really like to be able to do is show that the integral over Σ is negligible, so that we would have a surface integral just over the recording surface. This would be perfect for migration/inversion formulae. This issue is treated somewhat cavalierly in the literature. Σ is sent off to infinity and some argument about Sommerfeld is given. But for migration/inversion we want an anticausal Green function so as to propagate the recorded energy backwards in time. And, as we will now show, only the causal Green function satisfies the Sommerfeld condition as $r \to \infty$

As the observation point goes to infinity, the causal Green function has the form

$$G \to \frac{1}{4\pi R} e^{ikR} \tag{7.24}$$

Also, we can approximate $\frac{\partial G}{\partial R}$ by $\frac{\partial G}{\partial n}$. So as $R \to \infty$ we have

$$\frac{\partial G}{\partial n} \approx \frac{1}{4\pi R}\left[\frac{-1}{R}e^{ikR} + ike^{ikR}\right] = \frac{1}{4\pi R}e^{ikR}\left[ik - \frac{1}{R}\right] \tag{7.25}$$

and hence

$$\frac{\partial G}{\partial n} - i\frac{\omega}{c}G \approx \frac{K_1}{R^2} \tag{7.26}$$

as R goes to infinity. K_1 is just some constant factor. By the same argument

$$\frac{\partial u_S}{\partial n} - i\frac{\omega}{c}u_s \approx \frac{K_2}{R^2} \tag{7.27}$$

with a different constant factor K_2.

Now the integral whose value at infinity we need to know is

$$\int_S \left[u_S(\mathbf{x}, \mathbf{x}_S, \omega)\frac{\partial G(\mathbf{x}, \mathbf{r}, \omega)}{\partial n} - G(\mathbf{x}, \mathbf{r}, \omega)\frac{\partial u_S(\mathbf{x}, \mathbf{x}_S, \omega)}{\partial n}\right]dS. \tag{7.28}$$

If we add and subtract the term $i\omega/cGu_s$, we haven't changed anything, but we end up with

$$\int_S u_S\left[\frac{\partial G}{\partial R} - i\frac{\omega}{c}G\right] - G\left[\frac{\partial u_S}{\partial R} - i\frac{\omega}{c}u_s\right]dS. \tag{7.29}$$

The fields themselves (G and u_s) decay as one over R for large R. Therefore the complete integrand of this surface integral must decay as one over R^3. The surface area of S obviously increases as R^2. So we conclude that the surface integral itself must decay as one over R as R goes to infinity. In other words, if we use the causal Green function, we can neglect surface integrals of this form as the surface of integration goes to infinity.

But for migration we want an anti-causal Green function. And for the anti-causal Green function (G goes as e^{-ikR}/R) we have

$$\frac{\partial G}{\partial R} + i\frac{\omega}{c}G \approx \frac{K_1}{R^2} \tag{7.30}$$

where the + sign is especially important. Because of it $\frac{\partial G}{\partial R} - i\frac{\omega}{c}G$ must decay as one over R not R^2. The point is that we are stuck with using a causal expression for the scattered field since that is the result of a real physical experiment. But for the Green function we have a choice. Choosing the causal Green function results in an integrand which decays as one over R^3, making the surface integral go to zero as R goes to infinity. But choosing the anti-causal Green function results in an integrand which decays only as one over R^2.

In Schneider's original paper [41], he first did the argument with the causal Green fucntion, took the surface off to infinity and argued, quite properly, that this contribution from the surface at infinity was zero. But then when it came time to do migration, he slipped in an anti-causal Green function! As we have just seen this argument is false. So there is more to the question than this. Somehow we must show that the contribution to the scattered field from the integration over Σ is negligible compared to the integration over the recording surface S_0. The details of the argument are given in [18], but the upshot is that by using a stationary phase argument, one can indeed show that the integral over Σ is negligible.

8. The Method of Stationary Phase

We have various occasions in this course to consider approximating integrals of the form

$$I = \int_a^b f(x) e^{i\tau g(x)}\, dx \qquad (8.1)$$

where τ is a large parameter. Such integrals arise, for example, in the Green function integral representation of solutions of the wave equation in slowly varying media. Slowly varying material properties imply large relative wavenumbers. And in the chapter on phase-shift methods we saw that high-frequency solutions of the $c(z)$ wave equation can be written in the form

$$\psi = \frac{1}{\sqrt{k_z}} e^{\int k_z\, dz}.$$

There is a well-known method for treating integrals of this form, known as the method of stationary phase.[1] The basic idea is that as the parameter τ gets large, the integrand becomes highly oscillatory, effectively summing to zero: except where the phase function has a stationary point. We can illustrate this with a simple example shown in Figure 8.1. The top figure shows a Gaussian which we take to be the phase function $g(x)$. Clearly this has a stationary point at the origin. In the middle figure we show the real part of the hypothetical integrand at a relatively small value of τ. In the bottom figure, τ has been increased to 1000. You can see that only in the neighborhood of the origin is the integrand not highly oscillatory.

Stationary phase methods have been around since the 19th century. The name derives from the fact that the most significant contribution to I comes from those points where $g'(x) = 0$; i.e., the phase is stationary. We want an approximation that gets better as the parameter τ gets larger. If we integrate I by parts we get:

---

[1] This discussion follows that of Jones [32].

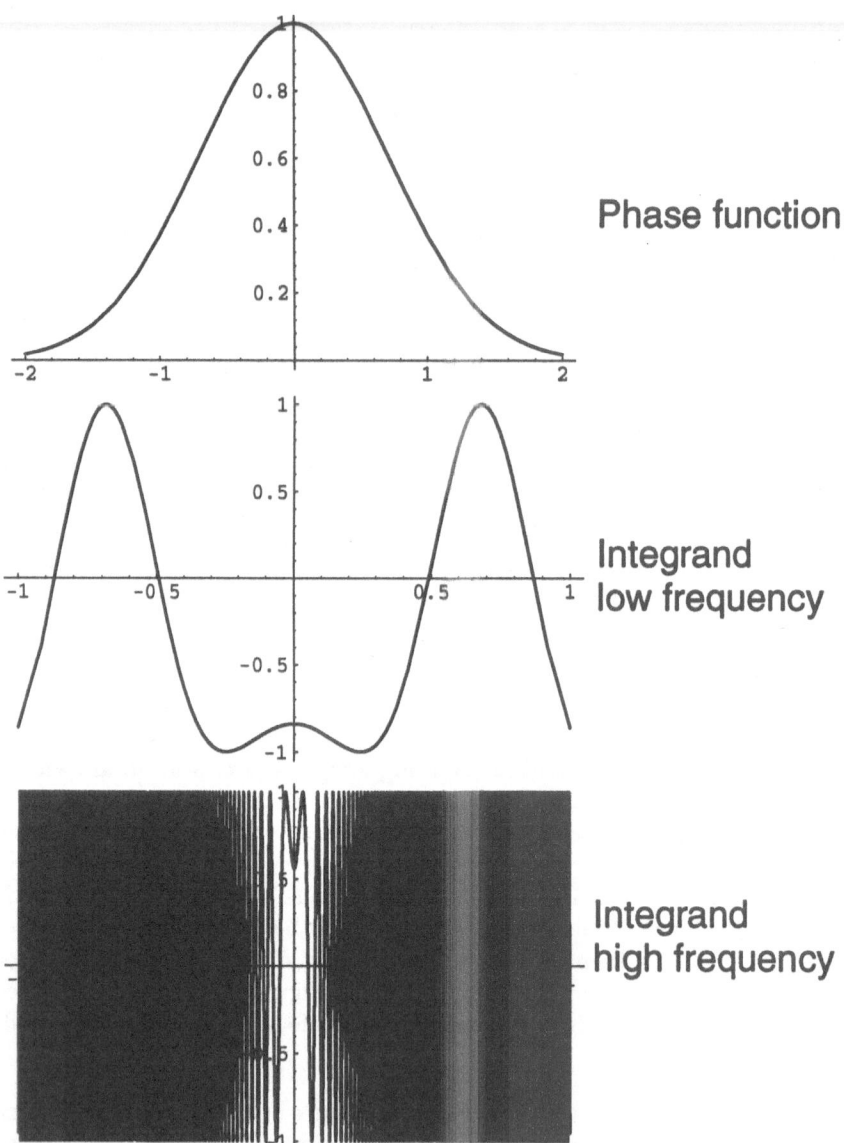

Phase function

Integrand
low frequency

Integrand
high frequency

Fig. 8.1. Hypothetical phase function (top); real part of the integrand $e^{i\tau g(x)}$ for $\tau = 10$ (middle); real part of the integrand for $\tau = 1000$.

$$I = \frac{f(x)e^{i\tau g(x)}}{i\tau g'(x)}\bigg|_a^b - \frac{1}{i\tau}\int_a^b \partial_x \left(\frac{f(x)}{g'(x)}\right) e^{i\tau g(x)}\, dx. \tag{8.2}$$

The first term is clearly $O(1/\tau)$. The second term will be $O(1/\tau^2)$ once we integrate by parts again–provided g' doesn't vanish in the interval $[a, b]$.

If g' is nonzero in $[a, b]$, we're done since as $\tau \to \infty$ we have the estimate

$$I \approx \frac{e^{i\tau g(x)}}{i\tau g'(x)}\bigg|_a^b. \tag{8.3}$$

But suppose $g'(x) = 0$ at some point x_0 in $[a, b]$. We can break op the integral I into three parts:

$$\int_a^{x_0-\eta} f(x)e^{i\tau g(x)}\, dx + \int_{x_0-\eta}^{x_0+\eta} f(x)e^{i\tau g(x)}\, dx + \int_{x_0+\eta}^b f(x)e^{i\tau g(x)}\, dx \tag{8.4}$$

where η is a small number. In the first and last of these integrals we can use the previous result since, by assumption, g is nonstationary in these intervals. The only question that remains is how the integral over the interval $[x_0 - \eta, x_0 + \eta]$ compares to these other contributions. Let's try approximating g by a Taylor series in the neighborhood of x_0:

$$g(x) = g(x_0) + g'(x_0)(x - x_0) + \frac{1}{2}g''(x_0)(x - x_0)^2 + \cdots \tag{8.5}$$

The second term in the Taylor series is zero by assumption. If we introduce the new variable $t = x - x_0$ then we have

$$\int_{x_0-\eta}^{x_0+\eta} f(x)e^{i\tau g(x)}\, dx = f(x_0) \int_{-\eta}^{\eta} e^{i\tau[g(x_0)+1/2g''(x_0)t^2]}\, dt. \tag{8.6}$$

We can pull the $f(x_0)$ out of the integral since η can be assumed to be as small as we like; this works provided f is not singular in the neighborhood of x_0. If it is, then we must resort to stronger remedies. Now if the functions f and g are well-behaved, then we can make the terms due to the integration by parts as small as we like since they are all at least $O(1/\tau)$. There are many complications lurking in the shadows and whole books are written on the subject (for example, [6]); but we will assume that we only have to worry about the contribution to I from the integral around the stationary point, Equation (8.6). In other words, we are making the approximation that we can write I as

$$I = f(x_0) \int_{-\infty}^{\infty} e^{i\tau[g(x_0)+1/2g''(x_0)t^2]}\, dt. \tag{8.7}$$

But this integral can be done analytically since

$$H = \int_{-\infty}^{\infty} e^{-x^2} \, dx$$

$$H^2 = \left[\int_{-\infty}^{\infty} e^{-x^2} \, dx \right] \left[\int_{-\infty}^{\infty} e^{-y^2} \, dy \right] = \int_{-\infty}^{\infty} \int_{-\infty}^{\infty} e^{x^2 + y^2} \, dx \, dy.$$

Therefore

$$H^2 = \int_{0}^{\infty} \int_{0}^{2\pi} e^{-r^2} r \, dr \, d\theta = \frac{1}{2} \int_{0}^{\infty} \int_{0}^{2\pi} e^{-\rho} \, d\rho \, d\theta = \pi$$

So $H = \sqrt{\pi}$ and

$$\int_{-\infty}^{\infty} e^{iat^2} \, dt = \sqrt{\frac{\pi}{a}} e^{i\pi/4}. \tag{8.8}$$

So we're left with

$$I \approx \sqrt{\frac{2\pi}{\tau g''(x_0)}} f(x_0) e^{i(\tau g(x_0) + \pi/4)}. \tag{8.9}$$

If g'' turns out to be negative, we can just replace g'' in this equation with $|g''| e^{i\pi}$.

If we're unlucky and g'' turns out to be zero at or very near to x_0 then we need to take more terms in the Taylor series. All these more sophistacated cases are treated in detail in *Asymptotic Expansions of Integrals* by Bleistein and Handelsman [6]. But this cursory excursion into asymptotics will suffice for our discussion of Kirchoff and WKBJ migration results.

Exercises

8.1 Consider the integral

$$I = \int_{a}^{b} f(x) e^{i\tau g(x)} \, dx.$$

Assuming that the phase $g(x)$ is nowhere stationary in the interval $[a, b]$, then I is $O(\tau^{-p})$ for large τ. Compute p.

8.2 If x_0 is a stationary point of g in the interval $[a, b]$ then

$$I \approx \sqrt{\frac{2\pi}{\tau g''(x_0)}} f(x_0) e^{i(\tau g(x_0) + \pi/4)}.$$

Show that this term dominates the end-point contribution, computed above, for large τ.

8.3 Use the method of stationary phase to approximate the integral

$$\int_{-\infty}^{\infty} e^{i\tau g(x)} \, dx$$

where $g(x) = e^{-t^2}$.

9. Downward Continuation of the Seismic Wavefield

9.1 A Little Fourier Theory of Waves

Just to repeat our sign conventions on Fourier transforms, we write

$$\psi(\mathbf{r}, t) = \int \int e^{i(\mathbf{k}\cdot\mathbf{r} - \omega t)} \psi(\mathbf{k}, \omega) \; d^3\mathbf{k} \; d\omega. \tag{9.1}$$

The differential volume in wavenumber space $d^3\mathbf{k} = dk_x \; dk_y \; dk_z$, is used for convenience. We're not going to use any special notation to distinguish functions from their Fourier transforms. It should be obvious from the context by simply looking at the arguments of the functions.

Now if the wavespeed in a medium is constant, then as with any constant coefficient differential equation, we assume a solution of exponential form. Plugging the plane wave $e^{i(\mathbf{k}\cdot\mathbf{r} - \omega t)}$ into the wave equation, $\nabla^2\psi - 1/c^2\partial_t^2\psi$, we see that our trial form does indeed satisfy this equation provided the wavenumber and frequency are connected by the dispersion relation: $k^2 \equiv \mathbf{k} \cdot \mathbf{k} = \omega^2/c^2$. Or, if you prefer to think of it in these terms, we have Fourier transformed the wave equation by replacing ∇ with $i\mathbf{k}$.

We can always simplify the analysis of the wave equation by Fourier transforming away one or more of the coordinates: namely any coordinates upon which the coefficients (in this case just c) do not depend. In exploration geophysics it's safe to assume the time-independence of c, so we can always make the replacement $\partial_t \rightarrow i\omega$. When it comes to the spatial variation of c we have to be more careful. After a constant velocity medium, the next simplest case to treat is a velocity which depends only on the depth z coordinate. If velocity (and hence the wavenumber) depends only on z, we can Fourier transform the x, y dependence of the Helmholtz equation $\nabla^2\psi + k^2\psi = 0$ to get

$$\frac{d^2\psi}{dz^2} + \left[k^2 - (k_x^2 + k_y^2) \right] \psi = 0. \tag{9.2}$$

Here we find that the trial solution

$$\psi = \psi_0 e^{\pm i k_z z} \tag{9.3}$$

works, provided that $k^2 = k_z^2 + k_x^2 + k_y^2$ as it must and $dk_z/dz = 0$. So how can we use this result in a $c(z)$ medium? There are two possibilities. The first (and most common in practice) assumption is that the velocity is actually piecewise constant; i.e., the medium is made up of a stack of horizontal layers, within each of which c is constant as shown in Figure 9.1. Or, put another way, if we let ψ_0

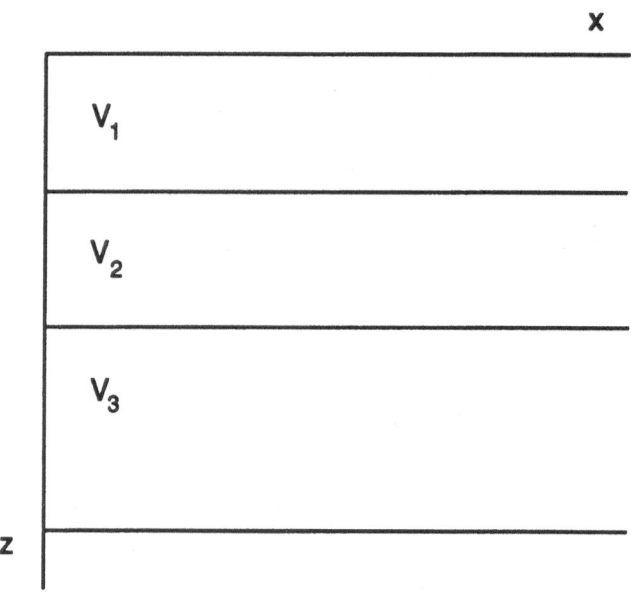

Fig. 9.1. A medium which is piecewise constant with depth.

denote the value of the wavefield at any depth z, then provided the velocity is constant in the strip $(z, z + \Delta z)$, the depth extrapolated wavefield at $z + \Delta z$ is

$$\psi = \psi_0 e^{i k_z |\Delta z|}. \tag{9.4}$$

Choice of the plus sign means that the plane wave

$$\psi = \psi_0 e^{i(k_z z - \omega t)} \tag{9.5}$$

will be a downgoing wave, because the phase stays constant if z is increased as t increases. To find the vertical wavenumber k_z we must Fourier transform the data and extract the horizontal wavenumbers k_x and k_y; k_z is then given by

$$k_z = \pm\sqrt{k^2 - k_x^2 - k_y^2}.$$

The other possibility for using the plane wave-like solutions in a $c(z)$ medium is within the WKBJ approximation.[1]

9.2 WKBJ Approximation

In fact we've already encountered the WKBJ approximation in another guise in the derivation of the eikonal equation, where we argued that the solution of the wave equation in a slowly varying medium could be represented in terms of slowly varying plane waves. Now we apply this idea to the $c(z)$ wave equation. The zero-th order approximation would be for k_z to be constant, in which case the solution to

$$\frac{d^2\psi}{dz^2} + k_z^2\psi = 0 \tag{9.6}$$

is just $\psi = e^{ik_z z}$.

Higher orders of approximation are achieved by making the slowly varying plane wave assumption: $\psi \approx e^{\phi(z)}$. (It doesn't matter whether we put an i in the exponent or not, it all works out in the end.) Then, in order for ψ to be a solution of Equation (9.6) it is necessary that

$$(\phi')^2 + \phi'' + k_z^2 = 0 \tag{9.7}$$

where primes denote differentiation with respect to z. The first order of approximation is to neglect the ϕ'' term. Then we can integrate the ODE exactly to get

$$\phi' = ik_z \tag{9.8}$$

and hence

$$\psi = e^{i\int k_z(z)\, dz}. \tag{9.9}$$

The second order approximation is a little harder to get. Using $\phi' = ik_z$ we can write $\phi'' = ik_z'$. So the nonlinear ODE for ϕ is now

$$(\phi')^2 + ik_z' + k_z^2 = 0 \tag{9.10}$$

and therefore

---

[1] The letters stand for G. Wentzel, H. Kramers, L. Brillouin, and H. Jeffreys

$$\phi' = ik_z\sqrt{1 + i\frac{k'_z}{k_z^2}}. \tag{9.11}$$

If we make the slowly-varying-media approximation again, then the second term in the square root is small compared to the first and we can simplify the square root using a first order Taylor approximation:

$$\begin{aligned} \phi &= i\int k_z - \frac{1}{2}\frac{k'_z}{k_z}\, dz \\ &= i\int k_z\, dz - \frac{1}{2}\ln k_z. \end{aligned} \tag{9.12}$$

So the second order WKBJ solution is

$$\psi = \frac{1}{\sqrt{k_z}}e^{i\int k_z\, dz}. \tag{9.13}$$

The singularity at $k_z = 0$ is just a reflection of the breakdown of the approximation that we made to achieve the result. Putting this all together, we have the plane wave like WKBJ solution in the space-time domain:

$$\frac{1}{\sqrt{k_z}}e^{i(\mathbf{k}\cdot\mathbf{r}-\omega t)} \tag{9.14}$$

where $\mathbf{k} = \hat{\mathbf{x}}k_x + \hat{\mathbf{y}}k_y \pm \hat{\mathbf{z}}\int k_z\, dz$. As usual, we choose the $+$ sign for downgoing waves and the $-$ sign for upgoing waves. The fact that these two waves propagate through the medium independently, as if there were no reflections is the crux of the WKBJ or geometrical optics approximation. Physically, the WKBJ wave corresponds to the primary arrival, and is not the result of any internal reflections within the model. In exploration geophysics, amplitude corrections are usually made on other grounds than WKBJ theory, so you will often see the square root term neglected. The WKBJ solution is then just a pure phase-shift.

9.3 Phase-Shift Migration

We have already seen that for piecewise constant $c(z)$ media, downward continuation of the wavefield is achieved by applying a series of phase-shifts to the Fourier transformed data. This is the essence of Gazdag's phase-shift method [21], perhaps the most popular $c(z)$ method. But let's formally introduce the method in its most common guise: 2D time migration. Once again we introduce the migrated or vertical travel time $\tau = z/v$. Instead of downward extrapolating the wavefield a depth increment Δz we use a vertical time increment $\Delta\tau = \Delta z/c$. Then the downward extrapolation operator we derived in the last section becomes

$$exp\left[-i\omega\Delta\tau\sqrt{1-\left(\frac{ck_x}{\omega}\right)^2}\right]. \tag{9.15}$$

To find the migrated image at a finite $\tau = n\Delta\tau$, we apply this operator n times in succession. But unlike Kirchhoff migration, where we applied the imaging condition once and for all, when we downward extrapolate the wavefield we must image at every depth. Fortunately, selecting the exploding reflector imaging condition $t = 0$ can be done by summing over all frequencies.[2]

Phase-Shift Migration à la Claerbout

$\psi_s(k_x, \omega) = Fourier\ Transform(\psi_s(x, t))$
$for(\tau = 0; \tau < \tau_{max}; \tau = \tau + \Delta\tau)$ {
    $for\ all\ k_x$ {
        $initialize:\ Image(k_x, \tau) = 0$
        $for\ all\ \omega$ {
            $C = exp[-i\omega\Delta\tau\sqrt{1-\left(\frac{ck_x}{\omega}\right)^2}]$
            $Image(k_x, \tau) = Image(k_x, \tau) + C * \psi_s(k_x, \omega)$
        }
        $Image(x, \tau) = Fourier\,Transform(Image(k_x, \tau))$
    }
}

Fig. 9.2. Claerbout's version of Gazdag's $c(z)$ phase-shift algorithm

For time migration, the migration depth step $\Delta\tau$ is usually taken to be the same as the sampling interval. So it will usually be the case that the complex exponential which does the downward extrapolation need not be recomputed for each $\Delta\tau$. Pseudo-code for the phase-shift algorithm is given in Figure 9.2.

The main drawback to the phase-shift migration would appear to be the improper way in which amplitudes are treated at boundaries. Rather than account for the partitioning of energy into reflected and transmitted waves at each boundary (as must surely be important if we have a layered model) we simply pass all the energy as if the transmission coefficient were unity. Presumably if the velocity is indeed slowly varying we would be better using a migration method based more explicitly on WKBJ approximations. The other possibility is to keep both the upgoing and downgoing solutions to the $c(z)$ wave equation, say with coefficients A^+ and A^-, and then apply the proper boundary conditions so that new A^+ and A^- are determined at each layer to conserve total energy. Classic phase-shift would then be the limiting case of setting the upgoing A coefficient to zero.

---

[2] If $f(t) = \int e^{i\omega t} f(\omega) d\,\omega$, then putting $t = 0$ gives $f(0) = \int f(\omega)\,d\omega$.

9.4 One-Way Wave Equations

When we can Fourier transform the wave equation, it is easy to select only downgoing wave propagation–we choose the positive square root in the dispersion relation. Unfortunately the Fourier transform approach is limited to homogeneous media. A technique to sort of reverse-engineer one-way wave propagation (suggested by Claerbout and described in [14], section 2.1) is to take an approximate dispersion equation, containing only downgoing wavenumbers, and see what sort of "wave equation" it gives rise to upon making the substitution: $ik_z \to \partial_z$, $ik_y \to \partial_y$, and so on. For example, start with the dispersion relation $k_z^2 + k_x^2 + k_y^2 = \frac{\omega^2}{c^2}$. Then solve for k_z and select downgoing wavenumbers by choosing the positive square root:

$$k_z = \frac{\omega}{c}\sqrt{1 - \frac{c^2 k_x^2}{\omega^2}} \tag{9.16}$$

and making the replacement $ik_z \to \partial_z$ we get the following downward continuation operator:

$$\partial_z \psi = i\frac{\omega}{c}\sqrt{1 - \frac{c^2 k_x^2}{\omega^2}}\,\psi \tag{9.17}$$

valid for depth-dependent media.

The problem with extending this to laterally varying media is the presence of the square root. One could imagine making the replacement $ik_x \to \partial_x$, but how do we take the square root of this operator? It is possible to take the square root of certain kinds of operators, but the approach most often taken in exploration geophysics is to approximate the square root operator, either by a truncated Taylor series or by a continued fraction expansion. The latter goes as follows.[3]

Write the dispersion relation

$$k_z = \frac{\omega}{c}\sqrt{1 - \frac{c^2 k_x^2}{\omega^2}} \tag{9.18}$$

as

$$k_z = \frac{\omega}{c}\sqrt{1 - X^2} = \frac{\omega}{c}R. \tag{9.19}$$

The n-th order continued fraction approximation will be written R_n and is given by

---

[3] For slightly more on continued fractions, see the section at the end of this chapter.

$$R_{n+1} = 1 - \frac{X^2}{1 + R_n}. \tag{9.20}$$

You can readily verify that this is the correct form of the approximation by noting that if it converges, it must converge to a number, say R_∞ which satisfies

$$R_\infty = 1 - \frac{X^2}{1 + R_\infty} \tag{9.21}$$

which means that $R^2 = 1 - X^2$.

The first few approximate dispersion relations using continued fraction approximations to the square root are given by:

$$k_z = \frac{\omega}{c} \qquad\qquad n=0 \tag{9.22}$$

$$k_z = \frac{\omega}{c} - \frac{ck_x^2}{2\omega} \qquad n=1 \tag{9.23}$$

$$k_z = \frac{\omega}{c} - \frac{k_x^2}{2\frac{\omega}{c} - \frac{ck_x^2}{2\omega}} \qquad n=2. \tag{9.24}$$

It is clear from the $n = 0$ term that these approximations rely to a greater (small n) or lesser (larger n) extent on the approximation that waves are propagating close to vertically. In fact it can be shown that the $n = 0$ term is accurate for propagation directions of within 5 degrees of vertical. The $n = 1$ term is accurate out to 15 degrees and the $n = 2$ term is accurate to 45 degrees. For more details on this aspect, see Claerbout ([14], Chapter 2.1). The different accuries of these dispersion relations are shown in Figure 9.3, which was computed with the following *Mathematica* code:

```
r[0,x_] = 1; r[n_,x_] := 1 - x^2/(1+ r[n-1,x]);
```

where I have taken $k^2 = 1$ for convenience. For example,

$$r[3,x] = 1 - \frac{x^2}{2 - \frac{x^2}{2 - \frac{x^2}{2}}} \tag{9.25}$$

To make use of these various approximations to the square root for $c(z,x)$ media, we proceed in two steps. First, since we are always going to be extrapolating downward in depth, so that we can assume that the velocity is constant in a thin strip Δz, we replace the vertical wavenumber k_z with $i\partial_z$. This gives us the following downward extrapolators according to Equations (9.22) - (9.24):

$$-i\frac{\partial \psi}{\partial z} = \frac{\omega}{c}\psi \qquad n=0 \tag{9.26}$$

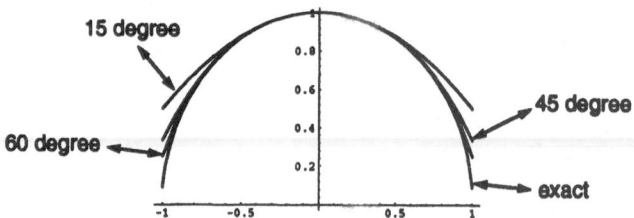

Fig. 9.3. The continued fraction approximation to the square root dispersion formula gives an approximate dispersion formula whose accuracy–reflected in the steepest angles at which waves can propagate–increases with the order of the approximation.

$$-i\frac{\partial\psi}{\partial z} = \left[\frac{\omega}{c} - \frac{ck_x^2}{2\omega}\right]\psi \qquad n=1 \qquad (9.27)$$

$$-i\frac{\partial\psi}{\partial z} = \left[\frac{\omega}{c} - \frac{k_x^2}{2\frac{\omega}{c} - \frac{ck_x^2}{2\omega}}\right]\psi \qquad n=2. \qquad (9.28)$$

Next, to account for the $c(x)$ variation, we replace k_x with $i\partial_x$. For example, using the $n = 1$ or 15 degree approximation we have:

$$\partial_z\psi = i\frac{\omega}{c}\left[1 + \frac{1}{2}\left(\frac{c}{\omega}\right)^2\partial_x^2\right]\psi. \qquad (9.29)$$

9.5 Paraxial Wave Equations

Yet another approach to achieving one-way propagation in laterally varying media is closely related to the WKBJ methods. In this approach we make the, by now standard, near-plane wave guess, but allow a laterally varying amplitude:

$$\psi = \chi(z,x)e^{i\omega(t-z/c)}. \qquad (9.30)$$

If we plug this form into the wave equation, what we discover is that in order for ψ to satisfy the wave equation it is necessary that χ satisfy

$$\nabla^2\chi + i\frac{2\omega}{c}\partial_z\chi = 0. \qquad (9.31)$$

So all we've done is rewrite the wave equation in terms of the new variable χ. The near-plane wave approximation comes in dropping the term $\partial_z^2\chi$ in comparison to $\partial_x^2\chi$. This results in a parabolic equation for χ:

$$\partial_z\chi = i\frac{c}{2\omega}\partial_x^2\chi. \qquad (9.32)$$

This is called a *paraxial* approximation since we are assuming that wave propagation occurs approximately along some axis, in this case the z axis. The paraxial
approximation is very popular in ray tracing since by making it we can generate
a whole family of rays in the neighborhood of a given ray for almost no extra
cost. In other words, we begin by tracing a ray accurately. Then, we generate a
family of paraxial rays in a neighborhood of this ray. How big this family can be
made depends on the extent to which we can drop the transverse Laplacian in
comparison with the axial Laplacian.

9.6 The Factored Two-Way Wave Equation

Another trick well-known from quantum mechanics is to factor the $c(z)$ wave
equation into an upgoing and a downgoing part. Clearly we can write

$$\frac{d^2\psi}{dz^2} + [k^2 - (k_x^2 + k_y^2)]\,\psi = \left[\frac{d}{dz} + ik\rho\right]\left[\frac{d}{dz} - ik\rho\right]\psi \qquad (9.33)$$

where $\rho = \sqrt{1 - (k_x^2 + k_y^2)/k^2}$.

If we denote the first differential operator in square brackets as L and the second
as L^*, then the $c(z)$ wave equation, Equation (9.2), which is $LL^*\psi = 0$, can be
written as a coupled first order system

$$\text{downgoing} \quad L^*\psi \;=\; \phi \qquad\qquad (9.34)$$
$$\text{upgoing} \quad L\phi \;=\; 0 \qquad\qquad (9.35)$$

As usual, for migration we neglect the coupling of upgoing and downgoing waves
and just consider the first order system describing downgoing waves.

If you're having trouble keeping straight which is the upgoing and which is the
downgoing equation, just remember what the two plane wave solutions of the
wave equation look like in the space-time domain: $e^{i(kz-\omega t)}$ and $e^{i(-kz-\omega t)}$. These
both satisfy the wave equation provided $k^2 = (\omega/c)^2$. In the first of these, z must
increase as t increases in order to keep the phase constant. In the second, z must
decrease as t increases in order to keep the phase constant. Therefore, $e^{i(kz-\omega t)}$
must be a downgoing wave and $e^{i(-kz-\omega t)}$ must be an upgoing wave.

9.7 Downward Continuation of Sources and Receivers

In this section we consider the non-zero-offset generalization of phase-shift migration. This is discussed in Section 3.3 of Claerbout's book [14]; we follow the simpler notation of [22] however.

Generally for 2D problems we have written the recorded data as: $\psi(x, z = 0, t)$. For slightly greater generality we can write this as $\psi(r, z_r, t)$ where z_r is the depth coordinate of the receiver. Implicitly the data also depend on the source coordinates, so we should really write $\psi(s, z_s; r, z_r)$, where z_s is the depth coordinate of the source. For a fixed source the wave equation is just

$$\partial_t^2 \psi = c^2 \left[\partial_r^2 \psi + \partial_{z_r}^2 \psi \right] . \tag{9.36}$$

The principle of reciprocity states that we should be able to interchange source and receiver coordinates and leave the observed wavefield unchanged. This is intuitively obvious on ray theoretic grounds; in practice it is very difficult to verify with field data since sources and receivers have different antenna characteristics. For example, it would seem that a weight drop should be reciprocal to a vertical geophone; a marine airgun reciprocal to a hydrophone. Claerbout says that reciprocity is verified in the lab to within the precision of the measurements and that in the field, small errors in positioning the sources and receivers can create discrepancies larger than the apparent discrepancies due to nonreciprocity. In any event, we will assume that reciprocity holds. That being the case we can write

$$\partial_t^2 \psi = c^2 \left[\partial_s^2 \psi + \partial_{z_s}^2 \psi \right] . \tag{9.37}$$

Assuming that the velocity depends only on depth, so that we can Fourier transform away the lateral coordinates, we end up with the two equations

$$\partial_{z_r}^2 \psi = - \left(\frac{\omega}{c} \right)^2 \left[1 - \left(\frac{k_r c}{\omega} \right)^2 \right] \psi \tag{9.38}$$

and

$$\partial_{z_s}^2 \psi = - \left(\frac{\omega}{c} \right)^2 \left[1 - \left(\frac{k_s c}{\omega} \right)^2 \right] \psi \tag{9.39}$$

where k_r and k_s are the horizontal wavenumbers in receiver and source coordinates.

Now these are the same $c(z)$ wave equations that we have dealt with many times before, just expressed in new coordinates. So we already know that both of these

equations have upgoing and downgoing solutions. With zero-offset data the source
and receiver coordinates are the same, so when we downward continued in the
exploding reflector model we only needed a single wave equation. But now we
want to allow for a greater degree of generality in order that we might be able to
apply the methods we've already discussed to far-offset data. So let us imagine
downward continuing both the sources and receivers simultaneously. At the reflector location the source and receiver coordinates must coincide. Coincident sources
and receivers imply that $t = 0$. This means that we can still use the exploding
reflector imaging condition. All we have to do is come up with one-way versions
of the **two** equations above. Clearly then we can use the one-way wave equations
that we've already derived. The choice of which one to use is ours. It is common,
however, to use the factored version of the wave equation, Equation (9.34). Then
we end up with

$$\partial_{z_r}\psi = \frac{i\omega}{c}\left[1 - \kappa_r^2\right]^{1/2}\psi \tag{9.40}$$

and

$$\partial_{z_s}\psi = \frac{i\omega}{c}\left[1 - \kappa_s^2\right]^{1/2}\psi \tag{9.41}$$

where $\kappa_r = ck_r/\omega$ and $\kappa_s = ck_s/\omega$.

When the source and receiver coordinates are coincident at the reflector, then
$z_r = z_s \equiv z$. Or, put another way, we can imagine lowering the sources and
receivers simultaneously. In either case

$$\partial_z\psi = \partial_{z_r}\psi + \partial_{z_s}\psi.$$

The way to think about this equation is that at a fixed receiver location, a small
change in the vertical location of a source or receiver is equivalent to a small
change in the observed travel times in the data. Therefore

$$\partial_z\psi = -\partial_z t \partial_t \psi$$

and so the total change in travel time associated with a perturbation dz of both
sources and receivers is given by

$$dt = [\partial_{z_r}t + \partial_{z_s}t]\,dz.$$

This allows us to write

$$\partial_z\psi = \frac{i\omega}{c}\left\{\left[1 - \kappa_s^2\right]^{1/2} + \left[1 - \kappa_r^2\right]^{1/2}\right\}\psi \tag{9.42}$$

which effects the simultaneous downward continuation of the sources and receivers referred to by Claerbout as *survey sinking*. Equation (9.42) is called the double-square-root equation.

The double-square-root equation is easily solved for $c(z)$ media–this is just our old friend the phase shift. Thus we can do nonzero-offset phase-shift migration since just as for the zero-offset case we can say that the data at any depth $z + \Delta z$ can be obtained from the data at z by

$$\psi(z + \Delta z) = \psi(z) e^{\frac{i\omega}{c}\left[(1-\kappa_s^2)^{1/2}+(1-\kappa_r^2)^{1/2}\right]\Delta z}. \tag{9.43}$$

We're almost finished now. All we have to do is apply the imaging condition. Part of this is just the usual exploding reflector $t = 0$, which we accomplish in the frequency domain by reintroducing ω and summing. The other part of the imaging condition is that $r = s$. This is conveniently expressed in terms of the half offset via $h = (r - s)/2 = 0$. So the fully migrated section is

$$\psi(x, h = 0, t = 0, z) = \sum_{k_x} \sum_{k_h} \sum_{\omega} \psi(k_x, k_h, \omega, z) e^{ik_s x} \tag{9.44}$$

where x is the midpoint $(r + s)/2$, $\kappa_r = (k_x + k_h)c/2\omega$ and $\kappa_s = (k_x - k_h)c/2\omega$. The summations over k_h and ω are the imaging conditions. The summation over k_x inverse Fourier transforms the wavefield back into midpoint space.

This downward continuation of sources and receivers, together with the zero-offset imaging condition, gives a prestack migration algorithm. Unfortunately it is of little utility as it stands because it requires that the velocity be laterally invariant: there's not much point in doing prestack migration if the velocity is really $c(z)$. One possible generalization would be to use a different velocity function for sources and receivers in Equation (9.40) and Equation (9.41). We can do this by downward continuing sources and receivers in alternating steps with the different velocity functions. The algorithm we end up with is:

– Sort data into common-source gathers $\psi(r, s_0, t, z)$ and extrapolate downward a distance Δz using Equation (9.41).

– Re-sort data into common-receiver gathers $\psi(r_0, s, t, z)$ and extrapolate downward a distance Δz using Equation (9.40).

– Compute the imaged CMP section by $m(x, z) = \psi(r = x, s = x, t = 0, z)$.

Clearly this method puts a tremendous burden on computer I/O if we're going to be continually transposing data back and forth between common-source and common-receiver domains. It might make more sense to simply migrate the individual common-source records. We can certainly downward continue a common source record using any of the one-way way equations that we've developed already. The only problem is what imaging condition do we use? The field scattered from any subsurface point (r', z') will image coherently at a time t' which equals the travel time from that point to the source point: $t'(s_0; r', s')$, where s_0 is a fixed source location. Here is the complete algorithm.

- Calculate the travel time $t'(s_0; r', s')$ from a given source location to all points in the subsurface.

- Downward continue (using Equation (9.41), for example) the shot record $\psi(r, s_0, t, z = 0)$. This results in the scattered field at each point in the subsurface $\psi(r, s_0, t, z)$.

- Apply the imaging condition $m(r', z'; s_0) = \psi(r', s_0, t'(s_0; r', s'), z')$.

If we used the correct velocity model to downward continue each shot record, then we should be able to sum all of the migrated common-source gathers and produce a coherent image of the subsurface. So we might write the final migrated image as

$$m(r', z') = \sum_s m(r', z'; s).$$

The main theoretical drawback of this approach to prestack shot migration is that it presupposes the existence of a single-valued travel time function. If, however, there is more than one ray between a source and any point in the subsurface, then there is no longer a well-defined imaging time. In this case we must resort to Claerbout's imaging condition, described by Paul Docherty in his lecture on Kirchhoff migration/inversion formulae.

9.8 Time Domain Methods

In the discussion of Kirchhoff migration methods we found that provided we had a Green function satisfying certain boundary conditions on the recording boundary $z = 0$, we could generate a solution of a boundary value problem for the wave equation anywhere in the subsurface and at all times. The boundary values that we use are the stacked seismic records. Therefore if we use an anticausal Green function, the solution of the wave equation that we get corresponds to the propagation of the zero-offset data backwards in time. Choosing $t = 0$ and $c = c/2$ yields an image of the subsurface reflector geometry. Further, the Green function solution is computed by integrating over the recording surface. Now since the time and the spatial coordinates are coupled via the travel time equation, this integration amounts to summing along diffraction hyperbolae. Nevertheless the form of the integration is that of a spatial convolution. We then saw that in the Fourier domain, this spatial convolution could be shown to be a phase-shift operation. And under the assumptions usually associated with time migration, we could think of this convolution as being a downward propagation of the recorded data.

Clearly there is a fundamental equivalence between propagating the recorded data backward in time and propagating it downward in depth. If the velocity model is constant or only depth dependent, then this equivalence is easily demonstrated by replacing time increments with migrated time or depth increments, suitably scaled by the velocity. In laterally heterogeneous media, the connection between the two is more complicated; in essence it's the wave equation itself which connects the two approaches.

When we discussed phase-shift migration methods, everything was done in the frequency domain, where it's difficult to see clearly how back-propagation in time is related. Downward propagation (at least for $v(z)$ media), is still a phase-shifting operation, however. So as we downward propagate the stacked seismic data from depth to depth, we have to inverse Fourier transform the results at each depth in order to apply the imaging condition. Fortunately, plucking off just the $t = 0$ value of the inverse transformed data is easy: just sum over frequencies. The point here is that we downward propagate the entire wavefield from depth to depth; at any given depth, applying the imaging condition results in a coherent image of that part of the wavefield which was reflected at that depth.

This all begs the question of why we can't simply use the time domain wave equation to backward propagate the stacked seismic data to $t = 0$, performing the imaging once and for all. We can. This sort of migration is called "reverse-time migration." But remember, in a fundamental sense all migration methods are reverse time. Certainly Kirchhoff is a reverse-time method, and as we have seen, by properly tracing the rays from source to receiver, we can apply Kirchhoff methods to complicated velocity models–certainly models where time migration is inappropriate. So the distinctions that are made in the exploration seismic literature about the various kinds of migration methods have little to do with the fundamental concepts involved and are primarily concerned with how the methods are implemented.

It appears that the first published account of what we now call reverse-time migration was given by Hemon [28]. His idea was to use the full wave equation, with zero "initial" values for ψ and $\partial_t \psi$ specified at some time T. The recorded data $\psi_s(x, t, z = 0)$ are then take as boundary values for a boundary value problem for the full wave equation, but in reverse time. This should cause no difficulty since the wave equation is invariant under a shift and reversal of the time coordinate $(t \to T - t)$.

The early papers on reverse time migration approximated the wave equation by an explicit finite difference scheme that was accurate to second order in both space and time. We will discuss such approximations in detail later, for now we simply exhibit such an approximation:

$$
\begin{aligned}
\psi(x_k, z_j, t_i) \;=\;& 2(1 - 2A^2)\psi(x_k, z_j, t_{i-1}) - \psi(x_k, z_j, t_{i-2}) \\
+\;& A^2 \left[\psi(x_{k+1}, z_j, t_{i-1}) + \psi(x_{k-1}, z_j, t_{i-1})\right. \\
+\;& \left.\psi(x_k, z_{j+1}, t_{i-1}) + \psi(x_k, z_{j-1}, t_{i-1})\right],
\end{aligned}
$$

where $A = c(x_k, z_j)\Delta t/h$, where h is the grid spacing in the x and z directions. This sort of approximation is called explicit because it gives the field at each grid-point at a given time-step in terms of the field at nearby gridpoints at the previous time-step: at each time-step t_i, everything to the right of the equal sign is known. Explicit finite difference methods are very easy to code and the above scheme can be used for both modeling and migration. In the latter case we simply run time backwards.

Initially, at the greatest time t_{max} recorded in the data, the field and its time derivative are set to zero in the entire subsurface. Then data at t_{max} provide values for the field at $z = 0$ and data at three times ending at t_{max} provide second order estimates of the time derivative of the field. The finite difference equation propagates these values into the subsurface for a single backward time-step $-\Delta t$ as data at $t_{max} - \Delta t$ are introduced at $z = 0$. We continue this procedure until $t = 0$. This has the effect of propagating the recorded data into the subsurface.

There are two extremely serious drawbacks with the finite difference approximation that we've introduced. The first is that low order schemes such as this require around 12 finite difference gridpoints per the shortest wavelength present in the data in order to avoid numerical dispersion. We can get around this problem by using a higher order scheme. A scheme which is accurate to fourth order in space and time requires only about 4 gridpoints per wavelength. The second problem is that using the full wave equation gives rise to spurious reflections at internal boundaries. These internal reflections are considered spurious because even if the model boundaries have been correctly located, the stacked data is presumed to have had multiples attenuated through the stacking procedure. The solution to this problem proposed by Baysal et al. [4] was to use a wave equation in which the density was "fudged" in order to make the normal-incidence impedance constant across boundaries. The idea being that since density is not something we normally measure in reflection seismology, we can make it whatever we want without affecting the velocities.

9.9 A Surfer's Guide to Continued Fractions

We illustrate here a few properties of continued fractions, adapted from the fascinating book *History of Continued Fraction and Padé Approximants* by Claude Brezinski [9].

Continued fractions represent optimal (in a sense to be described shortly) rational approximations to irrational numbers. Of course, we can also use this method to approximate rational numbers, but in that case the continued fraction expansion terminates after a finite number of terms. To get a feel for how these approximations work let's start with one of the oldest continued fractions; one which we can sum analytically. The first few terms are

$$S_0 = 0 \tag{9.45}$$

$$S_1 = \frac{1}{x} \tag{9.46}$$

$$S_2 = \frac{1}{x + \frac{1}{x}} \tag{9.47}$$

$$S_3 = \frac{1}{x + \frac{1}{x + \frac{1}{x}}} \tag{9.48}$$

$$S_4 = \frac{1}{x + \frac{1}{x + \frac{1}{x + \frac{1}{x}}}}. \tag{9.49}$$

The modern notation for this continued fraction expansion is

$$S = \frac{1|}{|x} + \frac{1|}{|x} + \frac{1|}{|x} \cdots \tag{9.50}$$

Each successive term S_n in this development is called a *convergent*. Clearly each convergent can, by appropriate simplification, be written as a rational number. For example,

$$S_2 = \frac{x}{x^2 + 1}$$

and

$$S_3 = \frac{1 + x^2}{2x + x^3}$$

It is easy to see that the following recursion holds:

$$S_2 = \frac{1}{x + S_1}, \; S_3 = \frac{1}{x + S_2}, \; S_4 = \frac{1}{x + S_3}\cdots \tag{9.51}$$

If this sequence converges then it must be that

$$S = \frac{1}{x + S} \tag{9.52}$$

which implies that

$$S^2 + xS - 1 = 0 \tag{9.53}$$

It is important to understand that we haven't actually proved that this continued fraction converges, we've simply shown that if it converges, it must converge to the quadratic in Equation (9.53). More generally, we can write the solution of the quadratic form

$$S^2 + bS - a = 0 \tag{9.54}$$

as

$$S = \frac{a|}{|b} + \frac{a|}{|b} + \frac{a|}{|b}\cdots \tag{9.55}$$

Some examples. Let $a = 1$ and $b = 1$ in (9.54). The first few convergents are

$$S_0 = 0, \; S_1 = 1, \; S_2 = \frac{1}{2}, \; S_3 = \frac{2}{3}, \; S_4 = \frac{3}{5}\cdots \tag{9.56}$$

This sequence converges to the positive root of the quadratic (9.54), namely $\frac{\sqrt{5}-1}{2}$, which was known as the *golden number* to the Greeks.[4] In fact the even and odd convergents provide, respectively, upper and lower bounds to this irrational number:

$$S_0 < S_2 < S_4 < \cdots \frac{\sqrt{5}-1}{2} \cdots < S_5 < S_3 < S_1 \tag{9.57}$$

Similarly, for $a = 1$ and $b = 2$, the even and odd convergents of 9.55 provide upper and lower bounds for the positive root of (9.54), namely $\sqrt{2} - 1$, to which they converge.

---

[4] *Mathematica* uses the golden number, or golden mean, as the default aspect ratio for all plots. To the Greeks, this number represented the most pleasing aspect, or proportion, for buildings and paintings. This view held sway well into the Renaissance.

This turns out to be a general result in fact: *the even/odd convergents of the continued fraction expansion of a number ξ always provide upper/lower bounds to ξ.* Also, one can show that the rational approximations to ξ generated by the convergents are optimal in the sense that the only rational numbers closer to ξ than a given convergent must have larger denominators.

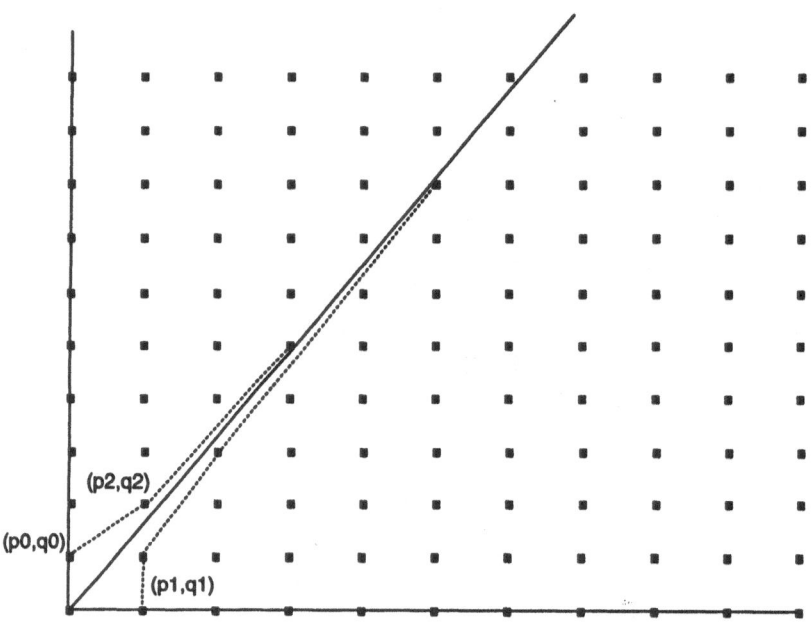

Fig. 9.4. An arbitrary number can be thought of as the slope of a line from the origin. The continued fraction approximation for the number (in this case the golden mean) consists of left and right polygons which approach an irrational number asymptotically and a rational number after a finite number of steps.

Since continued fractions are rational approximations we can give a nice geometrical illustration of what's going on in the approximation. Figure 9.4 shows the positive quadrant with a large dot at each integer. Each convergent, being a rational number, can be thought of as a line segment joining the dots. The asymptote of these successive line segments will be a line from the origin whose slope is the number we're approximating. If that number happens to be rational, the continued fraction series terminates, otherwise not. There are infinitely more irrational numbers than rational ones since the rationals are denumerable (can be put into one-to-one correspondence with the integers) while the irrationals are not. Therefore if we choose a number (or slope) at random, it is very unlikely that it will intersect one of the dots since that would imply that its slope were rational. This may appear counterintuitive, however, since it looks like there are an awful lot of dots!

In this particular example, the polygonal segments represent the convergents of the continued fraction for the Golden mean. The left and right vertices are given by the even and odd convergents

left: $p_0 = 0, q_0 = 1; p_2 = 1, q_2 = 2; p_4 = 3, q_4 = 5 \dots$

right: $p_1 = 1, q_1 = 1; p_3 = 2, q_3 = 3; p_5 = 5, q_5 = 8 \cdots$

There is a simple way to construct the continued fraction representation of any number ω. Begin by separating out the integral and decimal part of the number

$$\omega = n_0 + r_0 \qquad \text{where } 0 \le r_0 \le 1.$$

If the remainder r_0 is nonzero we repeat this procedure for $1/r_0$

$$1/r_0 = n_1 + r_1 \qquad \text{where } 0 \le r_1 \le 1.$$

We repeat this procedure until either it terminates with a zero remainder or we get tired. The result is

$$\omega = n_0 + \cfrac{1}{n_1 + \cfrac{1}{n_2 + \cfrac{1}{n_3 + \ddots}}}.$$

As an example, for π we get

$$\pi = 3 + \cfrac{1}{7 + \cfrac{1}{15 + \cfrac{1}{1 + \ddots}}}.$$

9.9.1 Fibonacci

The numbers appearing in the continued fraction expansion of the golden ratio might look familiar. They are also known as the Fibonacci numbers and they satisfy the recursion

$$F(0) = F(1) = 1; \quad F(j) = F(j-1) + F(j-2).$$

The Fibonacci numbers are named after Leonardo of Pisa, also known as Fibonacci (a contraction of *filius Bonacci*, son of Bonacci). Leonardo was born and died in Pisa, Italy, flourishing in the first part of the 13th century. He was a merchant who traveled widely in the Near East and became well acquainted with

Arabic mathematical work. The Fibonacci numbers were introduced in his study of the breeding habits of rabbits [9]. His *Liber abaci* was written in 1202 but not published until 1857.

As well as Fibonacci numbers, there are Fibonacci sequences. For example

$$S(0) = A;\ S(1) = B;\ S(j) = S(j-1), S(j-2).$$

So instead of adding the previous two iterates, we simply concatenate the previous two sequences. Fibonacci layered media are widely used to study disordered media, since although deterministic, they are very complex as the order of the sequence becomes large.

Exercises

9.1 Make the substitution $k_x \to i\partial_x$ and derive the 45 degree downward extrapolation equation. Hint: multiply both sides of the equation by the denominator of the continued fraction approximation so none of the derivatives are in the denominator.

9.2 Write your downward extrapolator as

$$\partial_z \psi = i\frac{\omega}{c}\psi + \text{another term.}$$

In optics these separate parts are known as, respectively, the *thin lens* term and the *diffraction* term. In practice, the downward extrapolation for $c(z)$ media might proceed in two phases for each Δz step: first we extrapolate using the thin lens term, then we apply the diffraction term.

9.3 Derive the time-domain version of your 45 degree equation by making the substitution $\omega \to i\partial_t$

9.4 Starting with the wave equation derive Gazdag's phase-shift migration method for $c(z)$ media.

9.5 Give a pseudo-code algorithm for phase-shift depth migration. What changes are necessary to make this a time migration algorithm? Under what circumstances would such a time migration algorithm be accurate.

9.6 The 45-degree approximate dispersion relation is

$$k_z = \frac{\omega}{c} - \frac{k_x^2}{2\frac{\omega}{c} - \frac{ck_x^2}{2\omega}}?$$

What is the (frequency-domain) wave equation associated with this dispersion relation.

9.7 If $\psi(r, s_0, t, z)$ is the downward continuation of the shot-record $\psi(r, s_0, t, 0)$, what is the migrated image associated with this shot record?

9.10 Computer Exercise: IV

Below is reproduced a complete migration code from the SU library written by John Stockwell and Dave Hale. Study the layout of this code and start thinking about how you will implement a straight-ray version of Kirchhoff migration. You should try to copy the outlines of this code (or other SU migration codes such as sumigtk.c and sustolt.c) insofar as the way parameters are read in, memory allocated, trace headers processed, and the final migrated section output. Do not reinvent these wheels, both because it would be a waste of time and also because the person who writes the best code will have the pleasure of seeing it added to the SU library. Try to be as modular in your writing as possible so that in a few weeks after we've talked about numerical methods for tracing rays you will be able to replace your straight ray calculation of travel times with a more sophisticated one. For this exercise, stick within the zero-offset approximation, but keep in mind how you might implement Claerbout's prestack imaging condition.

As a first step, write a Kirchhoff migration code that will use a completely general $c(x, z)$ velocity model, but don't worry about tracing rays to compute travel times. For this exercise you may either compute travel times along straight rays, or simply use the travel time associated with the rms velocity between the source and receiver. It would probably be best to assume the velocity model is defined on a grid. That should smoothly pave the way for the next exercise, which will involve accurate travel time calculations in moderately complex structures.

```
/* SUGAZMIG: $Revision: 1.6$;$Date: 92/10/28 16:09:01 $ */

/*-----------------------------------------------------------
 * Copyright (c) Colorado School of Mines, 1990.
 * All rights reserved.
 *
 * This code is part of SU.  SU stands for Seismic Unix, a
 * processing line developed at the Colorado School of
 * Mines, partially based on Stanford Exploration Project
 * (SEP) software.  Inquiries should be addressed to:
 *
 *   Jack K. Cohen, Center for Wave Phenomena,
 *   Colorado School of Mines,
 *   Golden, CO 80401
 *   (jkc@dix.mines.colorado.edu)
```

```
    *------------------------------------------------------------
    */

#include "su.h"
#include "segy.h"
#include "header.h"

/************** self documentation *********************/
char *sdoc[] = {
" ",
" SUGAZMIG - SU version of GAZDAG's phase-shift migration",
"      for zero-offset data. ",
" ",
" sugazmig <infile >outfile vfile= [optional parameters]",
" ",
" Optional Parameters: ",
" dt=from header(dt) or .004  time sampling interval ",
" dx=from header(d2) or 1.0   midpoint sampling interval",
" ft=0.0 first time sample ",
" ntau=nt(from data) number of migrated time samples ",
" dtau=dt(from header) migrated time sampling interval ",
" ftau=ft first migrated time sample ",
" tmig=0.0 times corresponding to interval velocities ",
in vmig",
" vmig=1500.0 interval velocities corresponding to times ",
in tmig",
" vfile= name of file containing velocities ",
" ",
" Note: ray bending effects not accounted for in this ",
version. ",
" ",
" The tmig and vmig arrays specify an interval velocity ",
" function of time. Linear interpolation and constant ",
" extrapolation is used to determine interval velocities ",
" at times not specified.  Values specified in tmig must ",
" increase monotonically.",
" ",
" Alternatively, interval velocities may be stored in",
" a binary file containing one velocity for every time",
" sample in the data that is to be migrated.  If vfile",
" is specified, then the tmig and vmig arrays are ",
" ignored.",
" ",
NULL};
```

```
/********* end self doc ******************************/

/*
 * Credits: CWP John Stockwell 12 Oct 1992
 *   Based on a constant v version by Dave Hale.
 *
 */

segy tr;

/* prototypes for functions defined and used below */
void gazdagvt (float k,
int nt, float dt, float ft,
int ntau, float dtau, float ftau,
float *vt, complex *p, complex *q);

/* the main program */
main (int argc, char **argv)
{
int nt; /* number of time samples */
int ntau; /* number of migrated time samples */
int nx; /* number of midpoints  */
int ik,ix,it,itau,itmig;/* loop counters  */
int nxfft; /* fft size */
int nk; /* number of wave numbers */
int np; /* number of data values */

int ntmig,nvmig;

float dt; /* time sampling interval  */
float ft; /* first time sample */
float dtau; /* migrated time sampling interval */
float ftau; /* first migrated time value  */
float dk; /* wave number sampling interval */
float fk; /* first wave number  */
float t,k; /* time,wave number */
float *tmig, *vmig; /* arrays of time, int. velocities */
float dx; /* spatial sampling interval */
float *vt; /* velocity v(t) */
float **p,**q; /* input, output data */

complex **cp,**cq; /* complex input,output */

char *vfile=""; /* name of file containing velocities */
```

```
    FILE *vfp; /* velocity file pointer */
    FILE *tracefp; /* fp for trace storage */
    FILE *headerfp; /* fp for header storage file    */

    /* hook up getpar to handle the parameters */
    initargs(argc,argv);
    requestdoc(1);

    /* get info from first trace */
    if (!gettr(&tr))  err("can't get first trace");
    nt = tr.ns;

    /* let user give dt and/or dx from command line */
    if (!getparfloat("dt", &dt)) {
    if (tr.dt) { /* is dt field set? */
    dt = (float) tr.dt / 1000000.0;
    } else { /* dt not set, assume 4 ms */
    dt = 0.004;
    warn("tr.dt not set, assuming dt=0.004");
    }
    }
    if (!getparfloat("dx",&dx)) {
    if (tr.d2) { /* is d2 field set? */
    dx = tr.d2;
    } else {
    dx = 1.0;
    warn("tr.d2 not set, assuming d2=1.0");
    }
    }


    /* get optional parameters */
    if (!getparfloat("ft",&ft)) ft = 0.0;
    if (!getparint("ntau",&ntau)) ntau = nt;
    if (!getparfloat("dtau",&dtau)) dtau = dt;
    if (!getparfloat("ftau",&ftau)) ftau = ft;

    /* store traces and headers in tempfiles while
    getting a count */

    tracefp = etmpfile();
    headerfp = etmpfile();
    nx = 0;
```

```
do {
 ++nx;
efwrite(&tr,HDRBYTES,1,headerfp);
efwrite(tr.data, FSIZE, nt, tracefp);
} while (gettr(&tr));
erewind(tracefp);
erewind(headerfp);

/* determine wavenumber sampling (for real
to complex FFT) */
nxfft = npfar(nx);
nk = nxfft/2+1;
dk = 2.0*PI/(nxfft*dx);
fk = 0.0;

/* allocate space */
p = alloc2float(nt,nxfft);
q = alloc2float(ntau,nxfft);
cp = alloc2complex(nt,nk);
cq = alloc2complex(ntau,nk);

/* load traces into the zero-offset array
and close tmpfile */
efread(*p, FSIZE, nt*nx, tracefp);
efclose(tracefp);

/* determine velocity function v(t) */
vt = ealloc1float(ntau);
if (!getparstring("vfile",&vfile)) {
ntmig = countparval("tmig");
if (ntmig==0) ntmig = 1;
tmig = ealloc1float(ntmig);
if (!getparfloat("tmig",tmig)) tmig[0] = 0.0;
nvmig = countparval("vmig");
if (nvmig==0) nvmig = 1;
if (nvmig!=ntmig) err("number tmig and vmig must be equal");
vmig = ealloc1float(nvmig);
if (!getparfloat("vmig",vmig)) vmig[0] = 1500.0;
for (itmig=1; itmig<ntmig; ++itmig)
if (tmig[itmig]<=tmig[itmig-1])
err("tmig must increase monotonically");
for (it=0,t=0.0; it<ntau; ++it,t+=dt)
intlin(ntmig,tmig,vmig,vmig[0],vmig[ntmig-1],
1,&t,&vt[it]);
```

```
} else {
if (fread(vt,sizeof(float),nt,fopen(vfile,"r"))!=nt)
err("cannot read %d velocities from file %s",nt,vfile);
}


/* pad with zeros and Fourier transform x to k */
for (ix=nx; ix<nxfft; ix++)
for (it=0; it<nt; it++)
p[ix][it] = 0.0;
pfa2rc(-1,2,nt,nxfft,p[0],cp[0]);


/* migrate each wavenumber */
for (ik=0,k=fk; ik<nk; ik++,k+=dk)
gazdagvt(k,nt,dt,ft,ntau,dtau,ftau,vt,cp[ik],cq[ik]);


/* Fourier transform k to x (including FFT scaling) */
pfa2cr(1,2,ntau,nxfft,cq[0],q[0]);
for (ix=0; ix<nx; ix++)
for (itau=0; itau<ntau; itau++)
q[ix][itau] /= nxfft;


/* restore header fields and write output */
for (ix=0; ix<nx; ++ix) {
efread(&tr,HDRBYTES,1,headerfp);
tr.ns = ntau ;
tr.dt = dtau * 1000000.0 ;
tr.delrt = ftau * 1000.0 ;
memcpy(tr.data,q[ix],ntau*FSIZE);
puttr(&tr);
}


}


void gazdagvt (float k,
int nt, float dt, float ft,
int ntau, float dtau, float ftau,
float *vt, complex *p, complex *q)
/*****************************************************
Gazdag's phase-shift zero-offset migration for one
wavenumber adapted to v(tau) velocity profile
*****************************************************
Input:
k wavenumber
```

```
nt number of time samples
dt time sampling interval
ft first time sample
ntau number of migrated time samples
dtau migrated time sampling interval
ftau first migrated time sample
vt velocity v[tau]
p array[nt] containing data to be migrated

Output:
q array[ntau] containing migrated data
*******************************************************
{
int ntfft,nw,it,itau,iw;
float dw,fw,tmax,w,tau,phase,coss;
complex cshift,*pp;

/* determine frequency sampling */
ntfft = npfa(nt);
nw = ntfft;
dw = 2.0*PI/(ntfft*dt);
fw = -PI/dt;

/* determine maximum time */
tmax = ft+(nt-1)*dt;

/* allocate workspace */
pp = alloc1complex(nw);

/* pad with zeros and Fourier transform t to w,
with w centered */
for (it=0; it<nt; it++)
pp[it] = (it%2 ? cneg(p[it]) : p[it]);
for (it=nt; it<ntfft; it++)
pp[it] = cmplx(0.0,0.0);
pfacc(1,ntfft,pp);

/* account for non-zero ft and non-zero ftau */
for (itau=0 ; itau < ftau ; itau++){
for (iw=0,w=fw; iw<nw; iw++,w+=dw) {
if (w==0.0) w = 1e-10/dt;
coss = 1.0-pow(0.5 * vt[itau] * k/w,2.0);
if (coss>=pow(ftau/tmax,2.0)) {
phase = w*(ft-ftau*sqrt(coss));
```

```
cshift = cmplx(cos(phase),sin(phase));
pp[iw] = cmul(pp[iw],cshift);
} else {
pp[iw] = cmplx(0.0,0.0);
}
}
}

/* loop over migrated times tau */
for (itau=0,tau=ftau; itau<ntau; itau++,tau+=dtau) {

/* initialize migrated sample */
q[itau] = cmplx(0.0,0.0);

/* loop over frequencies w */
for (iw=0,w=fw; iw<nw; iw++,w+=dw) {

/* accumulate image (summed over frequency) */
q[itau] = cadd(q[itau],pp[iw]);

/* compute cosine squared of propagation angle */
if (w==0.0) w = 1e-10/dt;
coss = 1.0-pow(0.5 * vt[itau] * k/w,2.0);

/* if wave could have been recorded in time */
if (coss>=pow(tau/tmax,2.0)) {

/* extrapolate down one migrated time step */
phase = -w*dtau*sqrt(coss);
cshift = cmplx(cos(phase),sin(phase));
pp[iw] = cmul(pp[iw],cshift);

/* else, if wave couldn't have been recorded in time */
} else {

/* zero the wave */
pp[iw] = cmplx(0.0,0.0);
}
}

/* scale accumulated image just as we would
for an FFT */
q[itau] = crmul(q[itau],1.0/nw);
}
```

```
/* free workspace */
free1complex(pp);
}
```

10. Plane Wave Decomposition of Seismograms

As we have seen throughout this course, wave equation calculations are greatly simplified if we can assume plane-wave solutions. This is especially true in exploration seismology since, to first order, the Earth is vertically stratified. Horizontal boundaries make Cartesian coordinates the natural choice for specifying boundary conditions. And if we apply separation of variables to the wave equation in Cartesian coordinates, we naturally get solutions which are represented as sums or integrals of plane waves.

One problem we face, however, is that plane waves are not physically realizable except asymptotically. If we are far enough away from a point source to be able to neglect the curvature of the wavefront, then the wavefront is approximately planar. The sources of energy used in exploration seismology are essentially point sources; but it is not clear when we can be considered to be in the far field.

On the other hand, plane waves, spherical waves, cylindrical waves, all represent "basis functions" in the space of solutions of the wave equation. This means that we can represent any solution of the wave equation as a summation or integral of these elementary wave types. In particular we can represent a spherical wave as a summation over plane waves. This means that we can take the results of a seismic experiment and numerically synthesize a plane wave response. We will begin this chapter by deriving the Weyl and Sommerfeld representations for an outgoing spherical wave. These are, respectively, the plane wave and cylindrical wave representations. Then, following the paper by Treitel, Gutowski and Wagner [48], we will examine the practical consequences of these integral relations. In particular we will examine the connection between plane wave decomposition and the slant-stack or $\tau - p$ transformation of seismology.

Migrating plane wave sections is no more difficult than migrating common-source records. The imaging condition is essentially the same, but expressed in terms of the plane wave propagation angle α. The comparison between shot migration and plane wave migration are nicely described in the paper by Temme [47]. Once we have the plane wave decomposition, we can migrate individual plane wave components. One reason this turns out to be important is that it seems that we can achieve high quality pre-stack migrations by migrating a relatively small number of plane wave components. In other words, plane wave migration appears to be more efficient than shot record migration since we can achieve good results with relatively few plane wave components. Common-source records, on the other hand, have a very limited spatial aperture.

10.1 The Weyl and Sommerfeld Integrals

The starting point for the plane wave decomposition is the Fourier transform of the spherical wave emanating from a point source at the origin:

$$\frac{e^{-ikr}}{r} = \int_{-\infty}^{\infty} \int_{-\infty}^{\infty} \int_{-\infty}^{\infty} A(k_x, k_y, k_z) e^{i(k_x x + k_y y + k_z z)} dk_x dk_y dk_z. \tag{10.1}$$

This equation looks like a plane wave decomposition already. It would appear that if we can invert the transform for the coefficients A, in other words, solve the integral

$$8\pi^3 A(k_x, k_y, k_z) = \int_{-\infty}^{\infty} \int_{-\infty}^{\infty} \int_{-\infty}^{\infty} \frac{e^{-ikr}}{r} e^{-i(k_x x + k_y y + k_z z)} dx \, dy \, dz \tag{10.2}$$

then we will be done. But look again. The integral for A involves all possible wave numbers k_x, k_y and k_z, not just those satisfying the plane wave dispersion relation. So in fact, Equation (10.2) is not a plane wave decomposition. We need to impose the dispersion relation as a constraint by eliminating one of the integrals over wave number. This is the approach taken in Chapter 6 of Aki and Richards [1], who evaluate the z integral using residue calculus and make the identification $k_z = \sqrt{k^2 - k_x^2 - k_y^2}$. If you are comfortable with residue calculus, then you should have a look at this derivation. Here we will follow a slightly simpler approach from [7]. We will first evaluate Equation (10.2) in the $x - y$ plane, then appeal to the uniqueness of solutions to the Helmholtz equation to extend the result to nonzero values of z. Many excellent textbooks exist which cover the use of complex variables in the evaluation of integrals, for example, Chapter 4 of Morse and Feshbach [37].

Let's begin by considering the special case $z = 0$. Then the problem is to evaluate

$$A(k_x, k_y) = \left(\frac{1}{2\pi}\right)^2 \int_{-\infty}^{\infty} \int_{-\infty}^{\infty} \frac{e^{-ikr}}{r} e^{-i(k_x x + k_y y)} dx dy. \tag{10.3}$$

We can evaluate this integral in polar coordinates. We need one set of angles for the wavevector $\mathbf{k}$ and another for the position vector $\mathbf{r}$: $k_x = q \cos \theta_2$, $k_y = q \sin \theta_2$, where $k_x^2 + k_y^2 = q^2$; and $x = r\cos\theta_1$, $y = r\sin\theta_1$. The element of area $dxdy = rdrd\theta_1$. So we have

$$k_x x + k_y y = rq[\cos\theta_2\cos\theta_1 + \sin\theta_2\sin\theta_1] = rq\cos(\theta_2 - \theta_1).$$

Therefore

$$A(k_x, k_y) = \left(\frac{1}{2\pi}\right)^2 \int_0^{2\pi}\int_0^{\infty} e^{-ir[k+q\cos(\theta_2-\theta_1)]}drd\theta_1 \tag{10.4}$$

The radial integral can be done by inspection

$$\int_0^{\infty} e^{-ir[k+q\cos(\theta_2-\theta_1)]}dr \equiv \int_0^{\infty} e^{-irB}dr = \left[\frac{irB}{iB}\right]_0^{\infty} \tag{10.5}$$

provided we can treat the contribution at infinity. Many books treat this sort of integral in an ad hoc way by supposing that there is a small amount of attenuation. This makes the wave number complex and guarantees the exponential decay of the anti-derivative at infinity ($k \to -ik'$ with $k' > 0$). If you want to look at it this way, fine. A more satisfying solution is to use the theory of analytic continuation. The integral in Equation (10.5) is the real part of the integral

$$\int_0^{\infty} e^{-irB'}dr \tag{10.6}$$

where B' is a complex variable $B + i\beta$. The integral clearly converges to $\frac{1}{iB'}$ provided $\beta < 0$. And this result is analytic everywhere in the complex plane except $B' = 0$. This means that we can extend, by analytic continuation, the integral in Equation (10.6) to the real B axis. In fact, we can extend it to the entire complex plane, except $B' = 0$. We can even extend it to zero values of B' by observing that this integral is really the Fourier transform of a step function. The result is that at the origin this integral is just pi times a delta function. The proof of this will be left as an exercise.

So we have

$$4\pi^2 A(k_x, k_y) = i\int_0^{2\pi}\frac{d\theta_1}{-[k+q\cos(\theta_2-\theta_1)]} = \frac{i}{k}\int_0^{2\pi}\frac{d\delta}{1+\frac{q}{k}\cos\delta} \tag{10.7}$$

where $\delta = \theta_2 - \theta_1$. The integral is now in a standard form:

$$\int_0^{2\pi}\frac{dx}{1+a\cos x} = \frac{2\pi}{\sqrt{1-a^2}}$$

if $a^2 < 1$. Putting this all together, we end up with

$$A(k_x, k_y) = \frac{i}{2\pi\sqrt{k^2 - k_x^2 - k_y^2}}$$

and hence

$$\frac{e^{ikr}}{r} = \frac{i}{2\pi}\int_{-\infty}^{\infty}\int_{-\infty}^{\infty}\frac{e^{i(k_x x + k_y y)}}{\sqrt{k^2 - k_x^2 - k_y^2}}dk_x dk_y. \tag{10.8}$$

This *is* the plane wave decomposition for $z = 0$ propagation. We can extend it to nonzero values of z in the following way. By inspection the general expression for this integral must have the form

$$\frac{e^{ikr}}{r} = \frac{i}{2\pi}\int_{-\infty}^{\infty}\int_{-\infty}^{\infty}\int_{-\infty}^{\infty}\frac{e^{i(k_x x + k_y y + k_z|z|)}}{k_z}dk_x dk_y dk_z \tag{10.9}$$

where we have made the identification $k_z = \sqrt{k^2 - k_x^2 - k_y^2}$.

The spherical wave e^{ikr}/r satisfies the Helmholtz equation for all z, the Sommerfeld radiation condition, and has certain boundary values at $z = 0$ given by Equation (10.8). You can readily verify that the integral representation, Equation (10.9), that we speculate is the continued version of Equation (10.8), also satisfies Helmholtz, Sommerfeld, and has the same boundary values on $z = 0$. It turns out that solutions of the Helmholtz equation which satisfy the Sommerfeld condition are unique. This means that Equation (10.9) must be the valid representation of e^{ikr}/r for nonzero values of z.

Equation (10.9) is know as the Weyl integral. We would now like to recast it as an integration over angles, to arrive at an *angular spectrum* of plane waves. In spherical coordinates, one has (cf. Figure 10.1)

$$k_x = k\cos\theta_2\sin\phi_2, \quad k_y = k\sin\theta_2\sin\phi_2, \quad k_z = k\cos\phi_2.$$

We have no problem letting θ_2 run from 0 to 2π. But a moment's reflection will convince you that we can't restrict ϕ_2 to real angles. The reason for this is that as k_x or k_y go to infinity, as they surely must since we are integrating over the whole $k_x - k_y$ plane, k_z must become complex. But

$$\cos\phi_2 = \frac{k_z}{k}.$$

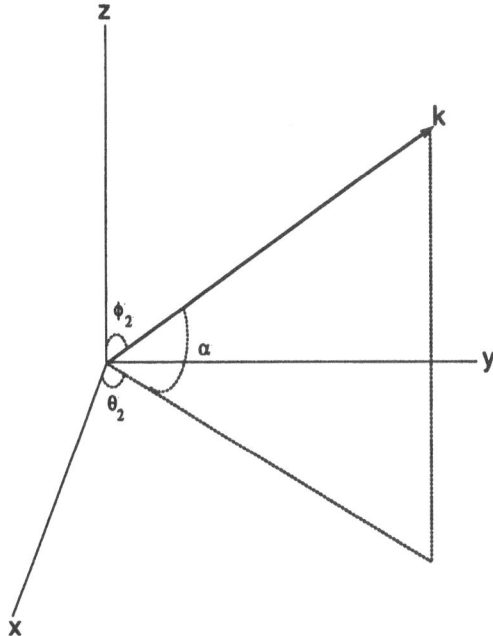

Fig. 10.1. Spherical coordinates representation of the wave vector.

Thus we see that **complex k_z values imply complex plane wave angles.** And if the angle of propagation of a plane wave is complex, it must decay (or grow) exponentially with distance. These sorts of waves are called evanescent or inhomogeneous waves. This is an extremely important point. The physical explanation given in Treitel et al. [48] is that these evanescent waves are source generated and that provided we are well away from the source we can ignore them. Thus in [48], the evanescent "waves" are ignored and the resulting integration over the angle ϕ_2 is restricted to the real axis. But remember, plane waves are nonphysical–they do not exist in nature! They are purely mathematical abstractions. Plane waves can be approximated in nature by various procedures. For example we could place a large number of point sources along a line and set them off together. This would generate a downgoing plane wave. Or we could apply a slight time delay between each source to simulate a plane wave incident at some angle. Another possibility is to be so far removed from a point source that the wavefront curvature is negligible.

A point source on the other hand is easily realizable: drop a pebble into a pond if you want to see one. So it should come as no surprise that we will have difficulty interpreting the results of a decomposition of a physical wave into non-physical waves. We have already seen that essentially any transient pulse can be synthesized from elementary Fourier components, but these don't exist in nature either. Any real signal must have a beginning. Remember also, that evanescent plane waves are not really waves at all. They do not suffer any phase shift upon translation and therefore have no travel time delay associated with them. This causes no practical problem so long as the origin of these terms is far enough away from any boundary that the exponential decay will have killed off any energy associated with them.

As to the direction of decay of these evanescent plane waves, that is completely arbitrary. In order to apply the dispersion relation as a constraint we needed to integrate out one of the wavevector components. We chose k_z because we like to think in terms of vertically layered media, where z is the naturally favored coordinate. The result of this was exponential decay of the inhomogeneous components in the z direction. But clearly we could have chosen any direction in k space and ended up with an analogous result.

The net result of all of this is that when k_x or k_y go to infinity, $k_z = i\infty$ which implies that $\cos\phi_2 = i\infty$. Now, using the fact that

$$\cos(\frac{\pi}{2} - ix) = \sin(ix) = i\sinh(x) = \frac{i}{2}(e^x - e^{-x})$$

it follows that when $\cos\phi_2 = i\infty$, $\phi_2 = \pi/2 - i\infty$. So that in spherical coordinates

$$\frac{e^{ikr}}{r} = \frac{ik}{2\pi} \int_0^{\pi/2-i\infty} \int_0^{2\pi} e^{i(k_x x + k_y y + k_z |z|)} \sin\phi_2 d\phi_2 d\theta_2. \qquad (10.10)$$

This first integral above is now taken to be a contour integral in the complex ϕ_2 plane. We integrate along the $\mathrm{Re}(\phi_2)$ axis from 0 to 2π; then we integrate to infinity in the direction of negative $\mathrm{Im}(\phi_2)$.

Finally, there is one other very useful form for the Weyl integral. It is achieved by introducing the radial wave number $k_r = k\cos\alpha$, where α is the angle of emergence shown in Figure 10.1. Then, using an integral representation of the cylindrical Bessel function (cf. [37]) the Weyl integral becomes

$$\frac{e^{ikr}}{r} = \frac{1}{2\pi} \int_0^\infty \int_0^{2\pi} e^{i(rk_r \cos(\theta_2 - \theta_1))} e^{-z\sqrt{k_r^2 - k^2}} \frac{k_r dk_r d\theta_2}{\sqrt{k_r^2 - k^2}} \qquad (10.11)$$

$$= \int_0^\infty \frac{J_0(k_r r) e^{-z\sqrt{k_r^2 - k^2}} k_r dk_r}{\sqrt{k_r^2 - k^2}}. \qquad (10.12)$$

In this form, the integral is known as the Sommerfeld integral and represents a sum over cylindrical waves. The cylindrical Bessel functions arise from the separation of variables of the wave equation in either cylindrical or spherical coordinates.

10.1.1 A Few Words on Bessel Functions

In any of the standard books on special functions, or what used to be called the functions of mathematical physics, you will see the following solutions of Bessel's equation:

$$J_\nu(x) = \left(\frac{x}{2}\right)^\nu \sum_{j=0}^\infty \frac{(-1)^j}{j!\Gamma(j+\nu+1)} \left(\frac{x}{2}\right)^{2j}$$

$$J_{-\nu}(x) = \left(\frac{x}{2}\right)^{-\nu} \sum_{j=0}^\infty \frac{(-1)^j}{j!\Gamma(j-\nu+1)} \left(\frac{x}{2}\right)^{2j}.$$

The limiting forms of these functions are:

$$x \ll 1 \qquad J_\nu(x) \rightarrow \frac{1}{\Gamma(\nu+1)} \left(\frac{x}{2}\right)^\nu$$

$$x \gg 1 \qquad J_\nu(x) \rightarrow \sqrt{\frac{2}{\pi x}} \cos(x - \nu\pi/2 - \pi/4).$$

The transition from the small argument behavior to the large argument behavior occurs in the region of $x \approx \nu$. Also notice that the large-argument asymptotic form demonstrates the existence of an infinite number of roots of the Bessel function. These are approximated asymptotically by the formula

$$x_{\nu n} \approx n\pi + (\nu - 1/2)\pi/2$$

where $x_{\nu n}$ is the n-th root of J_ν. You will find more than you ever wanted to know about Bessel Functions in Watson's *A Treatise on the Theory of Bessel Functions* [50]. Any of the classic textbooks on analysis from the early part of this century will have chapters on the various special functions. Morse and Feshbach [37] is another standard reference.

10.2 Reflection of a Spherical Wave from a Plane Boundary

The reason for going to all this trouble is so that we can treat the problem of the reflection of a pulse from a planar boundary. Clearly this is the crucial problem in reflection seismology. We can write the total potential as the incident plus reflected field

$$\psi = \frac{e^{ikr}}{r} + \psi_{\text{refl}}.$$

This is illustrated in Figure 10.2. A plane wave reflecting from a boundary produces a reflected plane wave. Therefore ψ must be representable as a superposition
of plane waves since the incident pulse is. All we need do is to take into account
the phase change associated with the travel time from the source location and the
boundary and to scale the amplitude of the reflected plane waves by the angle-
dependent reflection coefficient. Denoting the angle of incidence by α we have
from the Sommerfeld integral

$$
\begin{aligned}
\psi_{\text{refl}} &= \int_0^\infty \frac{J_0(k_r r) e^{-(h-z)\sqrt{k_r^2-k^2}} R(\alpha) k_r \, dk_r}{\sqrt{k_r^2-k^2}} \\
&= \int_0^\infty \frac{J_0(k_r r) e^{-i(h-z)k_z} R(\alpha)}{ik_z} k_r \, dk_r
\end{aligned}
\tag{10.13}
$$

where R is the reflection coefficient and h is the distance of the source above the
reflecting layer.

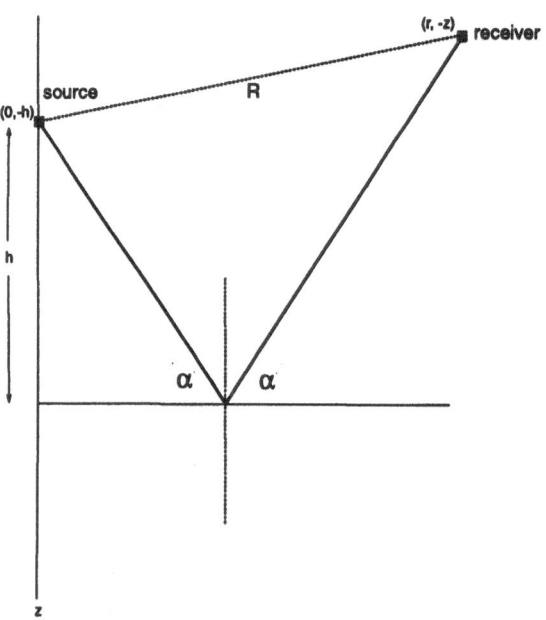

Fig. 10.2. Reflection of a pulse from a planar surface.

The plane wave incidence angle α is a function of both k_r and ω. In fact,

$$k_r = \frac{\omega}{c}\cos\alpha.$$

So we can write $R(\alpha) = \Phi(\omega, k_r)$. Further, so far we've neglected the frequency content of our point source, so if we want to interpret Equation (10.13) as, for example, the vertical p-wave displacement potential or pressure of a source plus reflected compressional wave, say $\psi(\omega, r, z)$, then we need to scale the integral by the Fourier transform of the source excitation function $F(\omega)$. This leads us to the first equation of Treitel et al. [48] and is the real jumping off point for the practical implementation of the plane wave decomposition of seismograms:

$$\psi(\omega, r, z) = F(\omega) \int_0^\infty \Phi(\omega, k_r) \frac{J_0(k_r r) e^{-i(h-z)k_z}}{i k_z} k_r dk_r. \tag{10.14}$$

Treitel et al. interpret $\Phi(\omega, k_r)$ as the Fourier transform of the velocity potential of the plane wave component with angular frequency ω and horizontal wave number k_r. For a single layer we know that in fact the reflection coefficient is independent of frequency. For a stack of layers it turns out that the effective reflection coefficient does indeed depend on the frequency. And in that case, we end up with a sum of phase shift terms associated with transmission of the plane wave through the layers above the reflecting layers. This is the basis of the well-known reflectivity method for synthetic seismogram generation [20]. Let's not worry about the source spectrum $F(\omega)$; for now let us assume that the source has a broad-band spectrum so that we can assume $F(\omega) = 1$. Then Equation (10.14) has the form of a Fourier-Bessel transform pair ([37], p. 766). For any function $g(r)$, we can define a Fourier-Bessel transform $G(k_r)$ via the invertible transformation pair

$$g(r) = \int_0^\infty G(k_r) J_0(k_r r) k_r dk_r$$

$$G(k_r) = \int_0^\infty g(r) J_0(k_r r) r dr.$$

This works just like a Fourier transform and with it we can invert the integral in Equation (10.14) analytically to achieve

$$\Phi(\omega, k_r) \frac{e^{-i(h-z)k_z}}{i k_z} = \int_0^\infty \psi(\omega, r, z) J_0(\frac{\omega r}{c}\sin\alpha) r\, dr. \tag{10.15}$$

For the moment let's not worry about the interpretation of Φ. If the medium is defined by a stack of horizontal layers, then we can calculate Φ straightforwardly using the methods in [20]. That's not the main goal of [48]; their main goal is to illuminate the connection between slant-stacks and plane wave decomposition. It will turn out that the slant-stacking procedure is essentially equivalent to plane wave decomposition. That means that instead of attempting to evaluate the Sommerfeld integral numerically, we can apply the slant-stacking procedure, which is widely used already in the exploration seismology community.

10.2.1 A Minor Digression on Slant-Stacks

Suppose we have some curve $t(x)$. For now, $t(x)$ is a purely geometrical object; we'll worry about the seismic implications later. Now, $t(x)$ *represents* an object, the curve. There is another, completely equivalent representation of this object, namely, as the envelope of all tangents to the curve (Figure 10.3). This equiva-

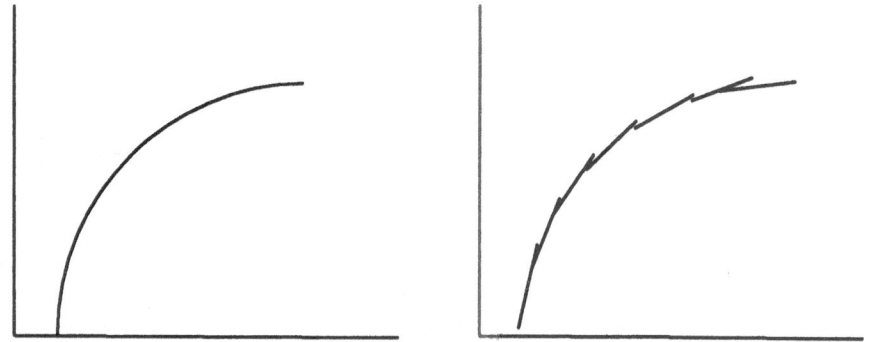

Fig. 10.3. A curve can be specified either as the locus of all points such that $t = t(x)$ or the envelope of all tangents with slope p and intercept τ.

lent geometry of slopes and intercepts is known as Pluecker *line geometry*. The conventional representation is known as *point geometry*. Line geometry forms the basis of the *Legendre* transformation, which is widely used in physics to switch between the Lagrangian and Hamiltonian pictures and from intensive to extensive variables. An excellent discussion of the Legendre transformation can be found in Callen's book *Thermodynamics* [10].

If we know the analytic form of a curve, it is straightforward to calculate its Legendre transformation into slope-intercept space. Let's consider a few examples. Since a straight line has only one slope and one intercept, the Legendre transformation of a line is a point. In other words, the line $t = px + \tau$ gets mapped into the point (τ, p).[1] Next, consider the quadratic $t = x^2$. The family of tangents to this curve must satisfy $t = px + \tau$ where $p = 2x$. In order that the lines $t = 2x^2 + \tau$ intersect the curve $t = x^2$ it is necessary that $\tau = -x^2$. Eliminating x between the expressions for τ and p we see that the quadratic $t = x^2$ gets mapped

---

[1] The reason for using p to represent the slope is that in seismology the slope is just the ray parameter, or horizontal phase slowness.

into the curve $\tau = -(p/2)^2$. Finally, consider the family of hyperbolae given by $t^2 = \alpha x^2 + \beta$. The family of tangents to these curves are given by $t = px + \tau$ where $p = \alpha x/\sqrt{\alpha x^2 + \beta}$. In order for these lines to intersect the hyperbolae it is necessary and sufficient that $\tau = \beta/\sqrt{\alpha x^2 + \beta}$. Eliminating x between the expressions for τ and p we obtain the result that the Legendre transformation of a $t - x$ hyperbola is a $\tau - p$ ellipse

$$\frac{p^2}{\alpha} + \frac{\tau^2}{\beta} = 1.$$

In the case of seismic data, we don't have an explicit relationship between the travel time t and the horizontal coordinate x or r. Of course, to the extent that we believe that diffraction signatures are hyperbolic, we should expect these to be mapped into $\tau - p$ ellipses. But we really need a numerical way to compute the $\tau - p$ transformation of seismic data. This is accomplished with the slant-stack: we sum along straight lines, each parameterized with a slope p and an intercept τ, and deposit the resulting value in the $\tau - p$ section. The reason this works is that unless the line we are summing along is approximately tangent to the curve, the summation results in a zero contribution. In other words, the intersection of the line of integration and the curve we wish to slant-stack is a set of measure zero and therefore contributes nothing to the slant-stack. On the other hand, a tangent has a "higher order contact" with a curve[2].

If we were simply dealing with plane curves which could be parameterized by $t(x)$, then we would have a complete picture: slant-stack equals Legendre transformation. With seismic data, the curve we are slant-stacking has some amplitude associated with it. So it sounds as if we should make the slant-stack a summation of amplitudes along lines. In the limiting case of an infinite frequency wavelet we just recover the *kinematical slant-stack* already discussed. In the more general case we have what is known as a *dynamical slant-stack*.

10.3 Slant-Stacks and the Plane Wave Decomposition

To show the connection between slant-stacks and the plane wave decomposition (following Treitel et al. [48]) we begin with an example of a slant-stack operator:

$$s(t, c_x) = \int_{-\infty}^{\infty} \int_{-\infty}^{\infty} u(x, y, t - x/c_x) dx \, dy \qquad (10.16)$$

---

[2] For the theory behind this, check out a book on differential geometry such as Struik's [45], where the result we need is on page 23.

where c_x is the horizontal phase velocity and u is our seismic data; it could be pressure or one component of displacement. We are clearly summing along lines of constant horizontal phase velocity and parameterizing the result by that phase velocity. Introducing polar coordinates

$$
\begin{aligned}
x &= r\cos\theta_1 \\
y &= r\sin\theta_1 \\
dx\,dy &= r\,dr\,d\theta_1
\end{aligned}
$$

we have

$$
s(t,\alpha) = \int_{-\pi}^{\pi}\int_0^{\infty} u(r, t - r\cos\theta_1\sin\alpha/c_r)r\,dr\,d\theta_1. \tag{10.17}
$$

or, since we are summing along a line, we can equivalently write this as a convolution with a delta function

$$
s(t,\alpha) = \int_{-\pi}^{\pi}\int_0^{\infty} u(r,t) * \delta(t - t_0\cos\theta_1)r\,dr\,d\theta_1 \tag{10.18}
$$

where $t_0 = r\sin\alpha/c$.

We can do the angular integral easily. Define the intermediate function $h(t)$ by

$$
h(t) = \int_{-\pi}^{\pi} \delta(t - t_0\cos\theta_1)d\theta_1 = 2\int_0^{\pi} \delta(t - t_0\cos\theta_1)d\theta_1.
$$

Letting $w = t_0\cos\theta_1$ we have

$$
h(t) = \int_{-t_0}^{t_0} (t_0^2 - w^2)^{-1/2}\delta(t - w)\,dw.
$$

The result is that

$$
h(t) = \begin{cases} 2(t_0^2 - t^2)^{-1/2} & \text{if } t \le t_0 \\ 0 & \text{otherwise} \end{cases} \tag{10.19}
$$

Thus, the slant-stack reduces to

$$
s(t,\alpha) = \int_0^{\infty} [u(t,r) * h(t,r,\alpha)]\,r\,dr. \tag{10.20}
$$

Believe it or not, the function we're convolving the data with, $h(t,r,\alpha)$, is a known Fourier transform pair of the 0-th order cylindrical Bessel function. One hesitates to say well-known, but in Oberhettinger's table of integrals (1957 edition, p. 61) we find that

$$h(t, r, \alpha) = \text{Fourier Transform} \left[2 \left[\left(\frac{r \sin \alpha}{c} \right)^2 - t^2 \right]^{-1/2} \right] \qquad (10.21)$$

$$= J_0 \left(\frac{\omega r}{c} \sin \alpha \right). \qquad (10.22)$$

We can now write the slant-stack in a form which is almost identical to the version of the plane wave decomposition we had in Equation (10.15):

$$s(\omega, \alpha) = \int_0^\infty u(\omega, r) J_0 \left(\frac{\omega r}{c} \sin \alpha \right) r \, dr. \qquad (10.23)$$

For comparison, we had for the plane wave decomposition of a pulse reflected from a horizontal boundary

$$\Phi(\omega, k_r) \frac{e^{-i(h-z)k_z}}{i k_z} = \int_0^\infty \psi(\omega, r, z) J_0 \left(\frac{\omega r}{c} \sin \alpha \right) r \, dr \qquad (10.24)$$

where $\psi(\omega, r, z)$ is the vertical p-wave displacement potential or pressure (i.e., e^{ikr}/r) and Φ was the angle-dependent reflection coefficient. The differences between the right sides of these two equations are a) that in the slant-stack we only know the field at the surface $z = 0$, whereas for the plane wave decomposition, the potential or pressure is assumed known everywhere is space; and b) if we interpret ψ as a potential, say the vertical p-wave displacement potential, then we have to differentiate it with respect to z in order to get the observed data u. Differentiation with respect to z just cancels the factor of ik_z in the denominator leaving

$$\Phi(\omega, k_r) e^{-i(h-z)k_z} = \int_0^\infty u(\omega, r, z) J_0 \left(\frac{\omega r}{c} \sin \alpha \right) r \, dr. \qquad (10.25)$$

We can interpret the phase-shift on the left side of this equation as an upward continuation operator. Once again, following the notation of [48], we suppose that the point source is buried at $(r = 0, z = h)$ and redefine the vertical distance between the point source and the receiver to be z. Then we have

$$\Phi(\omega, k_r) e^{-ik_z z} = \Phi(\omega, \alpha) e^{-iz\omega/c \cos \alpha} \qquad (10.26)$$

$$= \int_0^\infty u(\omega, r, z) J_0 \left(\frac{\omega r}{c} \sin \alpha \right) r \, dr. \qquad (10.27)$$

We'll simply define this to be $\Phi(\omega, \alpha, z)$. Then the plane wave component at $z = 0$, $\Phi(\omega, \alpha, 0)$ is given exactly by the slant-stack of the data:

$$\Phi(\omega, \alpha, 0) = \int_0^\infty u(\omega, r, 0) J_0\left(\frac{\omega r}{c} \sin \alpha\right) r \, dr. \tag{10.28}$$

Alternatively, starting with Equation (10.15)

$$\Phi(\omega, k_r) \frac{e^{-i(h-z)k_z}}{ik_z} = \int_0^\infty \psi(\omega, r, z) J_0(\frac{\omega r}{c} \sin \alpha) r \, dr. \tag{10.29}$$

we can differentiate with respect to z to get

$$\Phi(\omega, k_r) e^{-i(h-z)k_z} = \int_0^\infty u(\omega, r, z) J_0(\frac{\omega r}{c} \sin \alpha) r \, dr. \tag{10.30}$$

From this we infer that at the surface $z = 0$ we have

$$\Phi(\omega, k_r) e^{-ihk_z} = \int_0^\infty u(\omega, r, 0) J_0(\frac{\omega r}{c} \sin \alpha) r \, dr. \tag{10.31}$$

Or we could start directly with

$$\psi(\omega, r, z) = \int_0^\infty \frac{J_0(k_r r)}{ik_z} e^{-i(h-z)k_z} R(\alpha) k_r \, dk_r, \tag{10.32}$$

differentiate this with respect to z to get rid of the ik_z, and convert the potential into a displacement. Then, inverting the Fourier-Bessel transform we end up with

$$R(\alpha) e^{-ihk_z} \int_0^\infty u(\omega, r, 0) J_0(k_r r) r \, dr. \tag{10.33}$$

We've only just scratched the surface of plane-wave decompositions, reflectivity methods, and slant-stacks. In addition to the paper by Fuchs and Müller already cited [20], Chris Chapman has written a whole series of really outstanding papers on these subjects. Of special note are his papers on *Generalized Radon transforms and slant stacks* [13] and *A new method for computing synthetic seismograms* [12].

11. Numerical Methods for Tracing Rays

On page 89 we showed that the eikonal equation, which is a nonlinear partial differential equation, could be reduced to an ordinary differential equation by introducing new coordinates defined in terms of the surfaces of constant phase. The normals to these isophasal surfaces were called rays. Later we saw that the same ray equations followed automatically from Fermat's principle. Now we need to address the issue of how to solve these ray equations for realistic media.

There are essentially two classes of numerical methods for the ray equations. In the first, a heterogeneous medium (slowness as a function of space) is approximated by many pieces within each of which the ray equations are exactly solvable. For example, in a piece-wise constant slowness medium the exact solution of the ray equation must consist of a piece-wise linear ray, with Snell's law pertaining at the boundary between constant-slowness segments. One can show that in a medium in which the slowness has constant gradients, the ray equations have arcs of circles as solutions. So if we approximate our medium by one in which the slowness has constant gradients, then the raypath must be make up of lots of arcs of circles joined via Snell's law. For lack of a better term, let's call these methods algebraic methods. On the other hand, if we apply finite difference methods to the ray equations, we can simply regard the slowness as being specified on a grid and numerically determine the raypaths. Let's call these methods of solving the ray equations finite difference methods.

In practice, both algebraic and finite difference methods are applied to tomography, where raypaths must be known, and to migration, where we only need to know travel times. However, it is probably true that algebraic methods are more popular for tomography and finite difference methods are more popular for migration. One reason for this is that finite difference methods are easier to write; and since Fermat's principle guarantees that perturbations with respect to raypath are of higher order than perturbations with respect to the slowness, the inaccuracies in raypaths due to not applying Snell's law exactly at boundaries have a relatively minor effect on the quality of a migration. On the other hand, in travel time tomography, having accurate raypaths is sometimes vital for resolution.

11.1 Finite Difference Methods for Tracing Rays

We will now describe methods for solving the ray equations, Equation (11.1), when the slowness s is an arbitrary, differentiable function of x and z. The generalization to ray tracing in three dimensions is completely straight-forward. The methods to be discussed are classic and are described in much greater detail in the works of Stoer and Bulirsch [44] and Gear [23]. We will consider rays defined parametrically as a function of arc-length (or time or any other parameter which increases monotonically along the ray). In other words, a ray will be the locus of points $(x(\sigma), z(\sigma))$ as σ ranges over some interval. If we were to write the rays as simple functions $z(x)$ where x is the horizontal coordinate, then numerical difficulties would inevitably arise when the rays became very steep or turned back on themselves: if the ray becomes very steep, then finite differences $\Delta z / \Delta x$ may become unstable, forcing us to switch to a $x(z)$ representation, and if the ray turns back on itself, there will not be a single-valued representation of the ray at all.[1] Arc-length parameterization avoids all this difficulty, allowing one, for example, to decide on a step-size depending only on the intrinsic properties of the medium and not on the direction the ray happens to be propagating.

The first step is to reduce the second order Equation (11.1)

$$\frac{d}{d\sigma}\left[s(\mathbf{r})\frac{d}{d\sigma}\mathbf{r}\right] - \nabla s = 0 \tag{11.1}$$

to a coupled system of first order equations. The position vector $\mathbf{r}$ in Cartesian coordinates is simply $x\hat{\mathbf{x}} + z\hat{\mathbf{z}}$. Denoting differentiation with respect to the arc-length parameter σ by a dot one has

$$\frac{d\mathbf{r}}{d\sigma} = \dot{x}\hat{\mathbf{x}} + \dot{z}\hat{\mathbf{z}}.$$

---

[1] On the other hand, a $z(x)$ or $x(z)$ representation is certainly a little simpler, and results in a lower order system of equations to solve. To get the ray equations in this form, replace $d\sigma$ with

$$\sqrt{dx^2 + dz^2} = \sqrt{1 + (x'(z))^2}\ dz \text{ or } \sqrt{1 + (z'(x))^2}\ dx$$

in the travel time integral

$$t = \int_{\text{ray}} s(x, z)d\sigma$$

and proceed with the derivation of the Euler-Lagrange equations as we did at the end of the chapter on ray theory.

Carrying out the differentiations indicated in Equation (11.1) and equating the x and z components to zero we get

$$s\ddot{x} + s_x\dot{x}^2 + s_z\dot{z}\dot{x} - s_x = 0, \qquad (11.2)$$

and

$$s\ddot{z} + s_z\dot{z}^2 + s_x\dot{z}\dot{x} - s_z = 0. \qquad (11.3)$$

Now, to reduce a second order equation, say $y'' = f$, to two first order equations, it is necessary to introduce an auxiliary function, say z. Letting $y' = z$, the equation $y'' = f$ is completely equivalent to the two coupled equations

$$\begin{aligned} y' &= z \\ z' &= f. \end{aligned}$$

Carrying out this procedure with Equation (11.3), we arrive at the four coupled, first order, ordinary differential equations (ODE):

$$\begin{aligned} \dot{z}_1 &= z_2 \\ \dot{z}_2 &= -\frac{s_x z_2 + s_z z_4}{s}z_2 + \frac{s_x}{s} \\ \dot{z}_3 &= z_4 \\ \dot{z}_4 &= -\frac{s_z z_4 + s_x z_2}{s}z_4 + \frac{s_z}{s}, \end{aligned} \qquad (11.4)$$

where, $z_1 = x$, $\dot{z}_1 = \dot{x} = z_2$, and so on.

These are the coupled, nonlinear, ray equations in arc-length parameterization. We can simplify the notation by considering the four functions z_i, $i = 1, 2, 3, 4$, as being components of a vector $\mathbf{z}$, and the right hand side of Equation (11.4) as being the components of a vector $\mathbf{f}$. Then Equation (11.4) reduces to the single first order, vector ODE

$$\frac{d\mathbf{z}}{d\sigma} = \mathbf{f}(\mathbf{z}, s, s_x, s_z, \sigma).$$

Thus, we can dispense with vector notation and, without loss of generality, restrict attention to the general, first order ODE

$$\frac{dz}{dx} = f(x, z), \qquad (11.5)$$

where we will now use x as the independent variable, to conform with standard notation, and use primes to denote differentiation. The remainder of this chapter will be devoted to numerically solving initial value problems (IVP) and boundary value problems (BVP) for Equation (11.5).

11.2 Initial Value Problems

The initial value problem for Equation (11.5) consists in finding a function $z(x)$[2] which satisfies Equation (11.5) and also satisfies

$$z(x_0) = z_0. \tag{11.6}$$

It can be shown that Equations (11.5) and 11.6 have a unique solution provided that (a) f satisfies the Lipschitz[3] condition on the closed, finite interval $[a, b]$, (b) $z(x)$ is continuously differentiable for all x in $[a, b]$, (c) $z'(x) = f(x, z(x))$ for all x in $[a, b]$, and (d) $z(x_0) = z_0$. It follows from the mean value theorem of calculus that a sufficient condition for f to be Lipschitz is that all the partial derivatives of f with respect to z exist and be bounded and continuous.

In addition to being much easier to solve than boundary value problems, initial value problems have the important property that the solution *always* depends continuously on the initial value. In other words, initial value problems are "well posed." (For a complete discussion see [44].) The potential for ill-posedness will turn out to be a very serious problem for two-point boundary value problems.

The simplest solution method for an IVP is to approximate the derivative with a first-order forward difference, so that Equation (11.5) becomes

$$\frac{z(x + h) - z(x)}{h} \cong f(x, z(x))$$

where h is the step size. This gives $z(x + h) \cong z(x) + hf(x, z(x))$, which in turn gives Euler's method

---

[2] Remember, x and z no longer refer to the Cartesian coordinates. They are simply the independent and dependent variables of an abstract ODE.

[3] f is said to satisfy the Lipschitz condition if there exists a number N such that for all x, $|f(x, z_1) - f(x, z_2)| \le N|z_1 - z_2|$.

$$
\begin{aligned}
\eta_0 &= z_0 \\
\eta_{i+1} &= \eta_i + hf(x_i, \eta_i) && i = 0, 1, 2, \ldots \\
x_{i+1} &= x_i + h.
\end{aligned}
\tag{11.7}
$$

Thus, the approximate solution η consists of polygonal, or piece-wise linear, approximation to the curve $z(x)$. We start with an initial value of x and integrate using the recursion formula above for η until we reach the desired ending value of x. Euler's method is a special case of the general one-step method

$$
\begin{aligned}
\eta_0 &= z_0 \\
\eta_{i+1} &= \eta_i + h\phi(x_i, \eta_i, f, h) && i = 0, 1, 2, \ldots \\
x_{i+1} &= x_i + h.
\end{aligned}
$$

ϕ is the iteration function; in other words, ϕ is simply a rule which governs how the approximation at any given step depends on the approximations at previous steps. It can be determined in various ways: via quadrature rules, truncation of Taylor series, etc. For the Euler method ϕ is simply the right-hand-side of the ODE. The global discretization error of the method is $err(x, h) = \eta(x, h) - z(x)$. The method is convergent if the limit of $err(x, h_n)$ (where $h_n = (x - x_0)/n$) is zero as n goes to infinity. If $err(x, h_n)$ is $0(h_n^p)$ $(p > 0)$ the method is said to be of order p.

Perhaps the most widely used one-step methods are of the Runge-Kutta type. An example of a fourth order Runge-Kutta scheme is obtained by using the iteration function

$$
\phi(x, z, h) = 1/6[k_1 + 2k_2 + 2k_3 + k_4]
$$

where

$$
\begin{aligned}
k_1 &= f(x, z) \\
k_2 &= f(x + h/2, z + k_1 h/2) \\
k_3 &= f(x + h/2, z + k_2 h/2) \\
k_4 &= f(x + h, z + k_3 h).
\end{aligned}
\tag{11.8}
$$

One way to organize the myriad of one step methods is as numerical integration formulae. For example, the solution of the IVP

$$z'(x) = f(x) \qquad z(x_0) = z_0 \tag{11.9}$$

has the solution

$$z(x) = z(x_0) + \int_{x_0}^{x} f(t)dt. \tag{11.10}$$

The Runge-Kutta method in Equation (11.9) is simply Simpson's rule applied to Equation (11.10).

Thus far, we have given examples of single-step integration methods; the approximate solution at any given step depends only on the approximate solution at the previous step. The natural generalization of this is to allow the solution at the, say, kth step to depend on the solution at the previous r steps, where $r \geq 2$. To get the method started one must somehow compute r starting values $\eta_0, \eta_1, \ldots, \eta_{r-1}$. We will not consider multi-step methods in any further detail; the interested reader is referred to [44] and [23] and the references cited therein.

We have now developed sufficient machinery to attack the initial value problem of raytracing. By applying one of the single-step solvers such as the Runge-Kutta method, Equation (11.8), to the ray equations, Equation (11.4), we can compute the raypath for an arbitrarily complex medium. We can think of "shooting" a ray off from a given point in the medium, at a given takeoff angle with one of the axes, and tracking its progress with the ODE integrator. This turns out to be all we really need to do migration. We don't really care where a ray goes, we just want to know what travel time curves look like so we can do migration. On the other hand, if we are doing travel time tomography, then we need to know not where a given ray goes, but which ray (or rays) connects a fixed source and receiver. Thus we need to be able to solve the boundary value problem (BVP) for Equation (11.4).

11.3 Boundary Value Problems

The numerical solution of two-point BVP's for ordinary differential equations continues to be an active area of research. The classic reference in the field is by Keller [33]. The development given here will only scratch the surface; it is intended merely to get the interested reader started in tracing rays.

We now consider boundary value problems for the ray equations. The problem is to compute a "ray" $z(x)$ such that

$$\frac{dz}{dx} = f(x, z) \quad \text{subject to } z(a) = z_a \text{ and } z(b) = z_b. \tag{11.11}$$

The simplest solution to the BVP for rays is to shoot a ray and see where it ends up, i.e., evaluate $z(b)$. If it is not close enough to the desired receiver location, correct the take-off angle of the ray and shoot again. One may continue in this way until the ray ends up sufficiently close to the receiver. Not surprisingly this is called the *shooting method*. This idea is illustrated in Figure 11.1, which shows the geometry for a reflection seismic experiment.

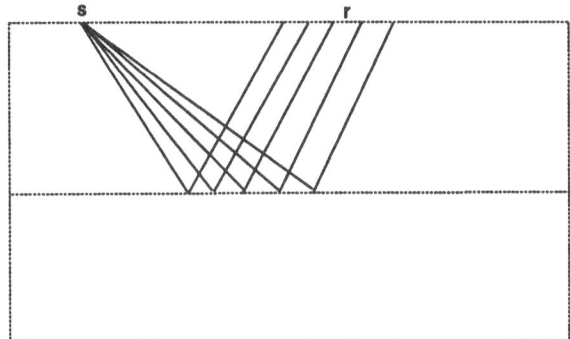

Fig. 11.1. Simple shooting involves solving the initial value problem repeatedly with different take-off angles, iteratively correcting the initial conditions until a satisfactory solution of the boundary value problem is achieved.

The method of correcting the take-off angle of subsequent rays can be made more rigorous as follows. By shooting, we replace the BVP Equation (11.10) with an IVP

$$\frac{dz}{dx} = f(x, z) \qquad z(a) = z_a \text{ and } z'(a) = \theta,$$

and adjust θ. Thus we can view a solution of this IVP as depending implicitly on θ, $z = z(x, \theta)$. The BVP then is reduced to the problem of finding roots of the function $z(b, \theta) - z_b$.

Having now reduced the problem to one of finding roots, we can apply the many powerful techniques available for this purpose, among which Newton's method is certainly one of the best. Newton's method is so important, underlying as it does virtually all methods for nonlinear inversion, that it merits a brief digression.

11.3.1 Newton's Method

Suppose you want to find roots of the function $f(x)$ (i.e., those points at which f is zero). If f is the derivative of some other function, say g, then the roots of f correspond to minima of g, so root finding and minimization are equivalent.

Let ζ denote the root of f to be computed. Expanding the function f about ζ in a Taylor series

$$f(\zeta) \equiv 0 = f(x) + f'(x)(\zeta - x) + \ldots,$$

we see that, neglecting higher order terms,

$$\zeta \cong x - \frac{f(x)}{f'(x)}.$$

To see intuitively what this expression means, consider the parabolic function shown in Figure 11.2. The tangent to the curve at $(x_0, f(x_0))$ satisfies the equation

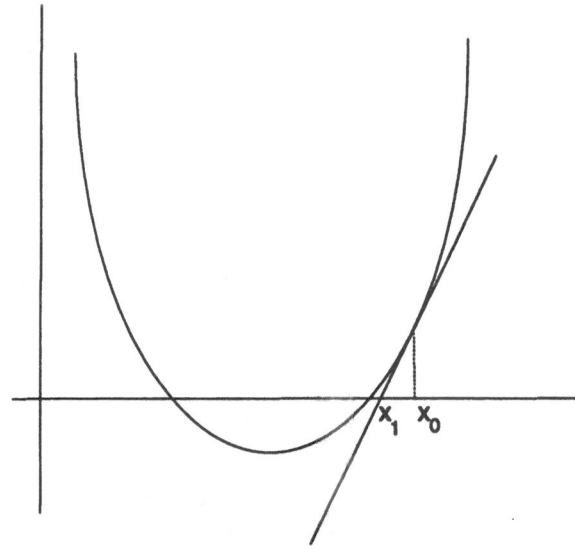

Fig. 11.2. Newton's method for finding roots.

$$y = f'(x_0)x_0 + \tilde{y}$$

where $\tilde{y}$ is the intercept with the y-axis. Eliminating $\tilde{y}$ from the following two simultaneous equations

$$f(x_0) = f'(x_0)x_0 + \tilde{y}$$
$$0 = f'(x_0)x_1 + \tilde{y}$$

gives

$$x_1 = x_0 - \frac{f(x_0)}{f'(x_0)}.$$

So we can interpret Newton's method as sliding down to the x-axis along a tangent to the curve associated with the initial point x_0. Thus, ζ will be closer to one of the two roots for any starting point we choose, save one. At the mid-way point between the two roots, the slope is zero and our approximation fails. Having computed ζ (call it ζ_0) we can insert it back into the above expression to arrive at the next approximation for the root

$$\zeta_1 \cong \zeta_0 - \frac{f(\zeta_0)}{f'(\zeta_0)}.$$

Continuing in this way we have the general first order Newton's method for finding the roots of an arbitrary differentiable function

$$\zeta_{n+1} \cong \zeta_n - \frac{f(\zeta_n)}{f'(\zeta_n)}.$$

Newton's method can be shown to be quadratically convergent, for suitably well-behaved functions f, provided the initial guess is close enough to the root ([44], p. 254 ff.).

Higher order Newton-type methods can be obtained by retaining more terms in the truncated Taylor series for f. For example, retaining only the linear terms amounts to approximating the function f with a straight line. Retaining the second order term amounts to approximating f with a quadratic, and so on. It is easy to see that the second order Newton method is just

$$\zeta_{n+1} \cong \zeta_n - \frac{f'(\zeta_n) \pm \left[f'(\zeta_n)^2 - 2f(\zeta_n)f''(\zeta_n)\right]^{1/2}}{f''(\zeta_n)}.$$

Newton's method has a very natural generalization to functions of more than one variable; but for this we need the definition of the derivative in R^n (n-dimensional Euclidean space). Writing the function $f(x)$, which maps from R^n into R^m, in terms of components as

$$f(x) = \begin{bmatrix} f_1(x_1, x_2, \ldots x_n) \\ f_2(x_1, x_2, \ldots x_n) \\ \vdots \\ f_m(x_1, x_2, \ldots x_n) \end{bmatrix},$$

the derivative of f is defined as the linear mapping D which satisfies

$$f(\zeta) = f(x) + Df(x)(\zeta - x),$$

in the limit as $x \to \zeta$. Defined in this way D is none other than the Jacobian of f

$$Df(x) = \begin{bmatrix} \dfrac{\partial f_1}{\partial x_1} & \cdots & \dfrac{\partial f_1}{\partial x_n} \\ & \vdots & \\ \dfrac{\partial f_m}{\partial x_1} & \cdots & \dfrac{\partial f_m}{\partial x_n} \end{bmatrix}.$$

So instead of the iteration

$$\zeta_{n+1} \cong \zeta_n - \frac{f(\zeta_n)}{f'(\zeta_n)},$$

we have

$$\zeta_{n+1} = \zeta_n - \Big(Df(\zeta_n)\Big)^{-1} f(\zeta_n),$$

provided, of course, that the Jacobian is invertible.

Example. Let's consider an example of Newton's method applied to a 1D function

$$q(x) = 5 - 6x^2 + x^4. \tag{11.12}$$

We have

$$q'(x) = -12x + 4x^3 \tag{11.13}$$

and therefore

$$\frac{q(x)}{q'(x)} = \frac{5 - 6x^2 + x^4}{-12x + 4x^3}. \tag{11.14}$$

Newton's iteration is then

$$x_{i+1} = x_i - \frac{5 - 6x_i^2 + x_i^4}{-12x_i + 4x_i^3} \tag{11.15}$$

The function q has four roots, at ± 1 and $\pm\sqrt{5}$. If we choose a starting point equal to, say, 3, we converge to $\sqrt{5}$ to an accuracty of one part in a million within 5 iterations. Now the point Newton's method converges to, if it converges at all, is not always obvious. To illustrate this, we show in Figure 11.3 a plot of the function q above a plot of the points to which each starting value (uniformly distributed along the abscissa) converged. In other words, in the bottom plot, there 600 tick-marks representing starting values distributed uniformly between -3 and 3 along the x-axis. The ordinate is the root to which each of these starting values converged after 20 iterations of Newton's method. Every starting point clearly converged to one of the four roots (zero is carefully excluded since q' is zero there). But the behavior is rather complicated. It's all easy enough to understand by looking at the tangent to the quadratic at each of these points; the important thing to keep in mind is that near critical points of the function, the tangent will be very nearly horizontal, and therefore a single Newton step can result in quite a large change in the approximation to the root. One solution to this dilemma is to use step-length relaxation. In other words, do not use a full Newton step for the early iterations. See [29] for details.

11.3.2 Variations on a Theme: Relaxed Newton Methods

The archetypal problem is to find the roots of the equation

$$g(x) = 0. \tag{11.16}$$

The most general case we will consider is a function g mapping from $\mathbf{R}^m$ to $\mathbf{R}^n$. Such a function has n components $g_i(x_1, x_2,, \ldots, x_m)$.

Newton's method can be written in the general case as the iteration

$$x_{k+1} = x_k - H_k g(x_k) \tag{11.17}$$

where H_k is the inverse of the Jacobian matrix $J(x_k)$. Now let us relax the strict definition of H and require that

$$\lim_{k \to \infty} \|I - H_k J(x_0)\| = 0. \tag{11.18}$$

Such an algorithm will have similar convergence properties as the full Newton's method. The details of this theory can be found in books on optimization, such as [29]. We give a few important examples here.

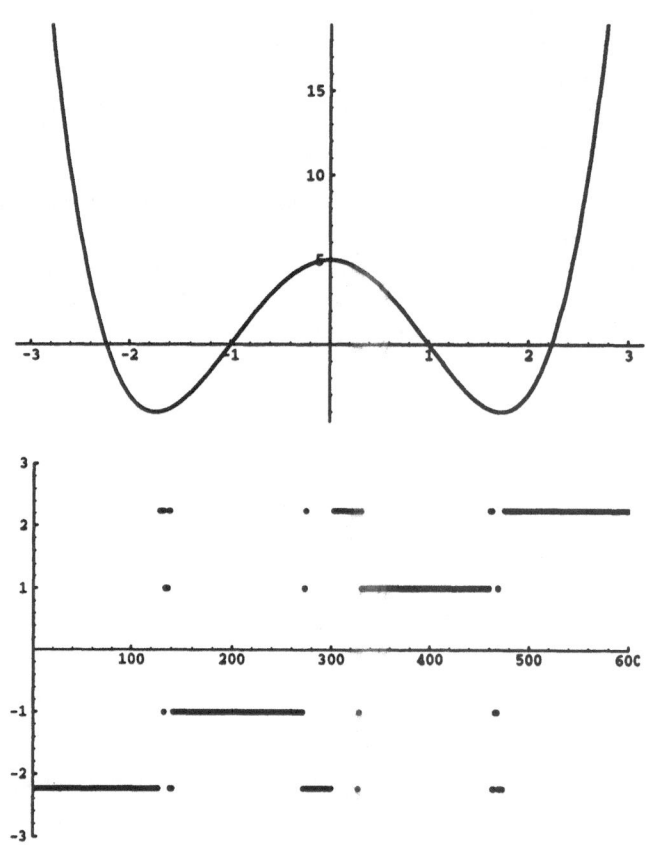

Fig. 11.3. Newton's method applied to a function with four roots, ± 1 and $\pm\sqrt{5}$. The bottom figure shows the roots to which each of 600 starting values, uniformly distributed between -3 and 3, converged. The behavior is complex in the vicinity of a critical point of the function, since the tangent is nearly horizontal.

- $H_k \to J^{-1}(x_0)$. An example of this would be step length relaxation. An example of step length relaxation on the function shown in Figure 11.3 is given in [40]. Here $H_k = a_k J^{-1}(x_0)$ where $a_k \to 1$.

- Use divided differences to approximate the derivatives. For example, with $n = 1$ we have,

$$x_{k+1} = x_k - \frac{(x_k - x_{k-1})g(x_k)}{g(x_k) - g(x_{k-1})}.$$

- Adjoint approximation. Take $H_k = a_k J^T(x_x)$. This approximation is extremely important in seismic inversion where it corresponds essentially to performing a prestack migration, given that g is a waveform misfit function. See [46] for more details.

11.3.3 Well-Posed Methods for Boundary Value Problems

Conceptually, the simplest method for solving two-point boundary value problems for the ray equations is shooting; we solve a sequence of initial value problems which come closer and closer to the solution of the boundary value problem. Unfortunately shooting can be ill-posed in the sense that for some models, minute changes in the takeoff angles of rays will cause large changes in the endpoints of those rays. This is especially true near "shadow zones". In fact, one is tempted to say that in practice, whenever a shooting code runs into trouble, it's likely to be due to the breakdown in geometrical optics in shadow zones. Put another way, the rays which cause shooting codes to break down almost always seem to be associated with very weak events.

On the other hand, there are many techniques for solving boundary value problems for rays which are not ill-posed and which will generate stable solutions even for very weak events in shadow zones. We will now briefly mention several of them. First there is multiple-shooting. This is an extension of ordinary shooting, but we choose n points along the x-axis and shoot a ray at each one–joining the individual rays continuously at the interior shooting points as illustrated in Figure 11.4.

Bending is an iterative approach to solving the ray equations based on starting with some initial approximation to the two-point raypath–perhaps a straight line– then perturbing the interior segments of the path according to some prescription. This prescription might be a minimization principle such as Fermat or it might employ the ray equation directly. One especially attractive idea along these lines is to apply a finite difference scheme to the ray equations. This will give a coupled set of nonlinear algebraic equations in place of the ODE. The boundary conditions can then be used to eliminate two of the unknowns. Then we can apply Newton's method to this nonlinear set of equations, the initial guess to the solution being a straight line joining the source and receiver. For more details see [44].

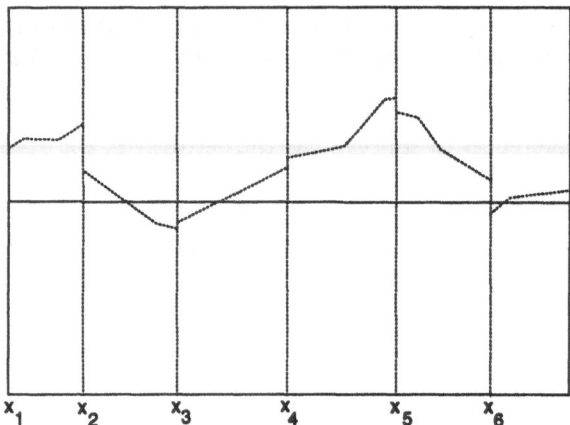

Fig. 11.4. Multiple-shooting involves shooting a number of rays and joining them continuously.

11.4 Algebraic Methods for Tracing Rays

If the slowness or velocity assume certain functional forms, then we can write down analytic solutions to the ray equation. Cerveny [11] gives analytic solutions for various media, including those with a constant gradient of velocity c, constant gradient of c^{-n} and $\ln c$, and constant gradient of slowness squared s^2. For example, if the squared slowness has constant gradient

$$s^2 = a_0 + a_1 x + a_2 z + a_3 y$$

then the raypaths are quadratic polynomials in arclength. This is probably the simplest analytic solution for an inhomogeneous medium. Similarly, if the velocity has a constant gradient, then the raypath is an arc of a circle. However, the coefficients of this circular representation are much more cumbersome than in the constant gradient of squared slowness model.

We can readily extend this sort of approach to media that satisfy these special properties in a piecewise fashion. For example, if the model is composed of blocks or cells within each of which the velocity has a constant gradient and across which continuity of velocity is maintained, then the analytic raypath will consist of smoothly joined arcs of circles. For more details of this approach see the paper by Langan, Lerche and Cutler [35] which influenced many people doing seismic tomography.

11.5 Distilled Wisdom

We have presented all of the theoretical tools necessary to write an efficient and accurate raytracer for heterogeneous media. However, there are many practical considerations which arise only from years of experience in attempting to apply these methods. There is probably no better source of this *distilled wisdom* than the short paper by Sam Gray entitled *Efficient traveltime calculations for Kirchhoff migration* [26]. Gray shows how interpolation of traveltime curves and downward continuation of rays can significantly reduce the computational costs of traveltime based migration methods.

11.6 Eikonal Equation Methods

During the past 4 or 5 years, there has been an explosion of interest in computing travel times by solving the eikonal equation directly. The main reason for this is that finite difference methods for the eikonal equation are well suited for both vector oriented machines such as the Cray 2/Cray XMP and massively parallel machines such as the CM-200/CM-5. In this brief discussion we will content our-selves to look at just one aspect of this problem, as described in the work of Vidale and Houston [49].

Suppose we want to solve the eikonal equation $\|\nabla t\|^2 = s^2$ on a square 2D grid of size h. Consider a single unit of this grid bounded by four points whose travel times are (going counter clockwise starting at the lower left) t_0, t_1, t_3, t_2. So points 0 and 3 are diagonally opposite. Approximating the travel times at grid points 1, 2, and 3 as first-order Taylor series about the point 0, we have

$$t_2 = t_0 + h\frac{\partial t}{\partial y}$$

$$t_1 = t_0 + h\frac{\partial t}{\partial x}$$

and

$$t_3 = t_0 + h\left(\frac{\partial t}{\partial x} + \frac{\partial t}{\partial y}\right).$$

Putting these together we have

$$2h\frac{\partial t}{\partial y} = t_3 + t_2 - t_1 - t_0$$

and

$$2h\frac{\partial t}{\partial x} = t_3 + t_1 - t_2 - t_0$$

which gives

$$4h^2 \left\{ \left(\frac{\partial t}{\partial x}\right)^2 + \left(\frac{\partial t}{\partial y}\right)^2 \right\} \approx (t_3 + t_1 - t_2 - t_0)^2 + (t_3 + t_2 - t_1 - t_0)^2.$$

This is our finite difference approximation to the right hand side of the eikonal equation. If you carry though the algebra you will see that most of the cross terms cancel, so that we are left with just $(t_3 - t_0)^2 + (t_1 - t_2)^2$. Now, the squared slowness we approximate by its average value at the 4 grid points

$$s \approx \bar{s} = \frac{1}{4}(s_0 + s_1 + s_2 + s_3).$$

The result is Vidale's approximation for the eikonal equation

$$(t_3 - t_0)^2 + (t_1 - t_2)^2 = 2\bar{s}^2 h^2$$

or

$$t_3 = t_0 + \sqrt{2\bar{s}^2 h^2 - (t_1 - t_2)}.$$

11.7 Computer Exercises: V

Now extend your Kirchhoff code from Exercise IV by accurately calculating travel times in a general $c(x, z)$ medium. Whether you do this by tracing rays or by solving the eikonal equation is up to you. At this point you should not be concerned about efficiency. Try to get something that works, then worry about making it fast.

12. Finite Difference Methods for Wave Propagation and Migration

The most widely used technique for modeling and migration in completely hetero-geneous media is the method of finite differences. We will consider finite differences applied to two generic equations: the parabolic equation

$$\frac{\partial u}{\partial t} = K \frac{\partial^2 u}{\partial x^2}$$

which arises in this context from Claerbout's paraxial approximation to the wave equation, and the hyperbolic wave equation

$$\frac{\partial^2 u}{\partial t^2} = c^2 \frac{\partial^2 u}{\partial x^2}$$

where K and c are functions of space but not of time or u. Our treatment of finite differences will be extremely terse. For more details, the reader is urged to consult the standard references, including Smith's *Numerical solution of partial differential equations: finite difference methods* [42] (from which much of the first part of this chapter is directly drawn), Mitchell's *Computational methods in partial differential equations* [36], and Richtmyer and Morton's *Difference methods for initial value problems*. There are countless fine references to the theory of partial differential equations, but an especially useful one (by virtue of its brevity, read-ability and rigor) is John's *Partial Differential Equations* [31]. In particular we have relied on John's discussion of characteristics and the domain of dependence of PDE's and their associated finite difference representations.

First a little notation and nomenclature. We use U to denote the exact solution of some PDE, u to denote the exact solution of the resulting difference equation and N to denote the numerical solution of the difference equation. Therefore the *total error* at a grid point i, j is

$$U_{i,j} - N_{i,j} = (U_{i,j} - u_{i,j}) + (u_{i,j} - N_{i,j}) \tag{12.1}$$

which is just the sum of the *discretization error* $U_{i,j} - u_{i,j}$ and the *global rounding error* $u_{i,j} - N_{i,j}$

Let $F_{i,j}(u)$ denote the difference equation at the i,j mesh point. The *local truncation error* is then just $F_{i,j}(U)$. If the local truncation error goes to zero as the mesh spacing goes to zero, the difference equation is said to be *consistent*.

Probably the most important result in numerical PDE's is called Lax's Theorem: If a linear finite difference equation is consistent with a well-posed linear initial value problem, then stability guarantees convergence. A problem is said to be well-posed if it admits a unique solution which depends continuously on the data. For more details see [38].

Ultimately all finite difference methods are based on Taylor series approximations of the solutions to the PDE. For example, let U be an arbitrarily differentiable function of one real variable, then including terms up to order three we have:

$$U(x+h) = U(x) + hU'(x) + 1/2h^2U''(x) + 1/6h^3U'''(x) \qquad (12.2)$$

and

$$U(x-h) = U(x) - hU'(x) + 1/2h^2U''(x) - 1/6h^3U'''(x). \qquad (12.3)$$

First adding, then subtracting these two equations we get, respectively

$$U(x+h) + U(x-h) = 2U(x) + h^2U''(x) \qquad (12.4)$$

and

$$U(x+h) - U(x-h) = 2hU'(x). \qquad (12.5)$$

Thus we have the following approximations for the first and second derivative of the function U:

$$U''(x) \approx \frac{U(x+h) - 2U(x) + U(x-h)}{h^2} \qquad (12.6)$$

and

$$U'(x) \approx \frac{U(x+h) - U(x-h)}{2h}. \qquad (12.7)$$

Both of these approximations are second-order accurate in the discretization, i.e., $O(h^2)$. This means that the error will be reduced by one-fourth if the discretization is reduced by one-half. Similarly, if in Equations (12.2-12.3) we ignore terms which are $O(h^2)$ and higher we end up with the following first-order forward and backward approximations to $U'(x)$:

$$U'(x) \approx \frac{U(x+h) - U(x)}{h} \tag{12.8}$$

and

$$U'(x) \approx \frac{U(x) - U(x-h)}{h}. \tag{12.9}$$

12.1 Parabolic Equations

In the first part of this chapter we will consider functions of one space and one time variable. This is purely a notational convenience. Where appropriate, we will give the more general formulae and stability conditions. The discretization of the space and time axes is: $x = i\delta x = ih$, $t = j\delta t = jk$, for integers i and j. Then we have $U(ih, jk) \equiv U_{i,j}$. The simplest finite difference approximation to the nondimensional (constant coefficient) heat equation

$$\frac{\partial u}{\partial t} = \frac{\partial^2 u}{\partial x^2}$$

is thus

$$\frac{u_{i,j+1} - u_{i,j}}{k} = \frac{u_{i+1,j} - 2u_{i,j} + u_{i-1,j}}{h^2}. \tag{12.10}$$

12.1.1 Explicit Time-Integration Schemes

Now suppose that we wish to solve an initial-boundary value problem for the heat equation. In other words we will specify $U(x,t)$ for x at the endpoints of an interval, say $[0,1]$, and $U(x,0)$ for all x. Physically, this corresponds to specifying an initial heat profile along a rod, and requiring the endpoints of the rod have fixed temperature, as if they were connected to a heat bath at constant temperature.

We can re-write Equation (12.10) so that everything involving the current time step j is on the right side of the equation and everything involving the new time step $j+1$ is on the left:

$$u_{i,j+1} = u_{i,j} + r[u_{i-1,j} - 2u_{i,j} + u_{i+1,j}] \tag{12.11}$$

where $r = \frac{k}{h^2} = \frac{\delta t}{\delta x^2}$.

As an illustration of this finite difference scheme, consider the following initial-boundary value problem for the heat equation:

$$
\begin{aligned}
\frac{\partial U}{\partial t} &= \frac{\partial^2 U}{\partial x^2} \\
U(x = 0, t) &= U(x = 1, t) = 0 \\
U(x, 0) &= \begin{cases} 2x & \text{if } 0 \le x \le 1/2 \\ 2(1 - x) & \text{if } 1/2 \le x \le 1 \end{cases}
\end{aligned}
$$

Figure 12.1 shows the solution computed for 10 time steps using an x discretization $\delta x = .1$ and a t discretization $\delta t = .001$, thus giving $r = .1$. The errors are negligible, except at the point $x = 1/2$ where the derivative of the initial data is discontinuous. On the other hand, if we increase δt to .005, so that $r = .5$, the errors become substantial and actually grow with time (Figure 12.2). To see just what is happening, let us write this explicit time integration as a linear operator. Equation (12.11) can be expressed as

$$
\begin{bmatrix} u_{1,j+1} \\ u_{2,j+1} \\ u_{3,j+1} \\ \vdots \\ u_{N-1,j+1} \end{bmatrix} = \tag{12.12}
$$

$$
\begin{bmatrix} (1 - 2r) & r & 0 & 0 & \cdots \\ r & (1 - 2r) & r & 0 & \cdots \\ 0 & r & (1 - 2r) & r & \cdots \\ \vdots & & & & \ddots \\ 0 & \cdots & 0 & r & (1 - 2r) \end{bmatrix} \begin{bmatrix} u_{1,j} \\ u_{2,j} \\ u_{3,j} \\ \vdots \\ u_{N-1,j} \end{bmatrix}
$$

or as

$$
\mathbf{u}_{j+1} = \mathbf{A} \cdot \mathbf{u}_j
$$

where we use a bold-face $\mathbf{u}$ to denote a vector associated with the spatial grid. The single subscript now denotes the time discretization. We have by induction

$$
\mathbf{u}_j = \mathbf{A}^j \cdot \mathbf{u}_0.
$$

To investigate the potential growth of errors, let's suppose that instead of specifying $\mathbf{u}_0$ exactly, we specify some $\mathbf{v}_0$ which differs slightly from the exact initial data. The error is then $\mathbf{e}_0 = \mathbf{u}_0 - \mathbf{v}_0$. Or, at the j-th time step

$$
\mathbf{e}_j = \mathbf{A}^j \cdot \mathbf{e}_0.
$$

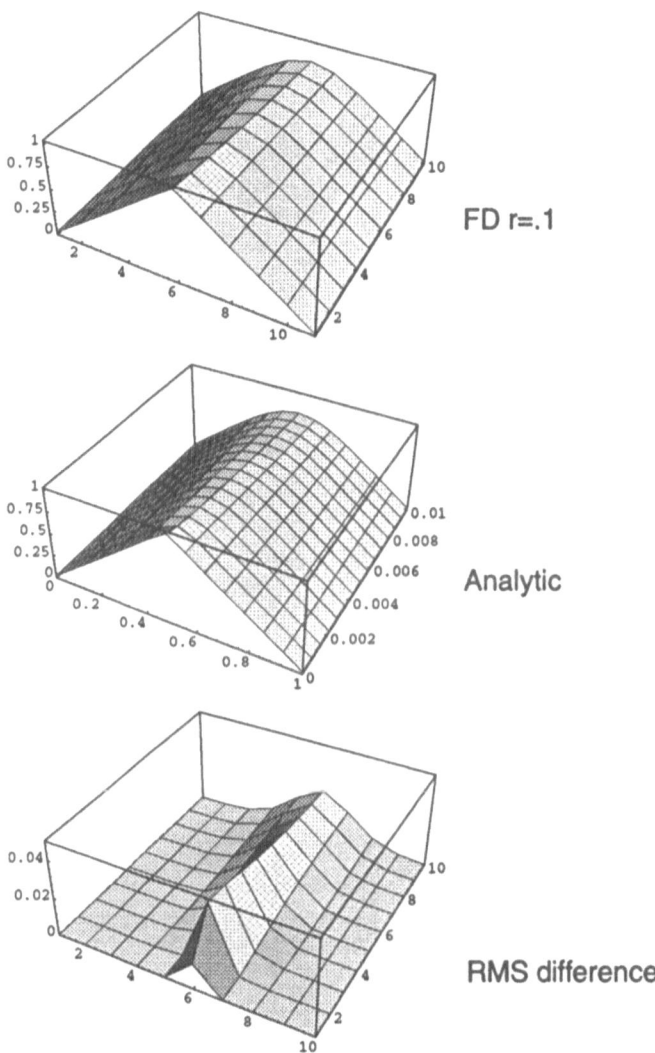

Fig. 12.1. Finite difference and analytic solution to the initial-boundary value problem for the heat equation described in the text. The discretization factor $r = .1$. The errors are negligible, except at the point $x = 1/2$ where the derivative of the initial data is discontinuous.

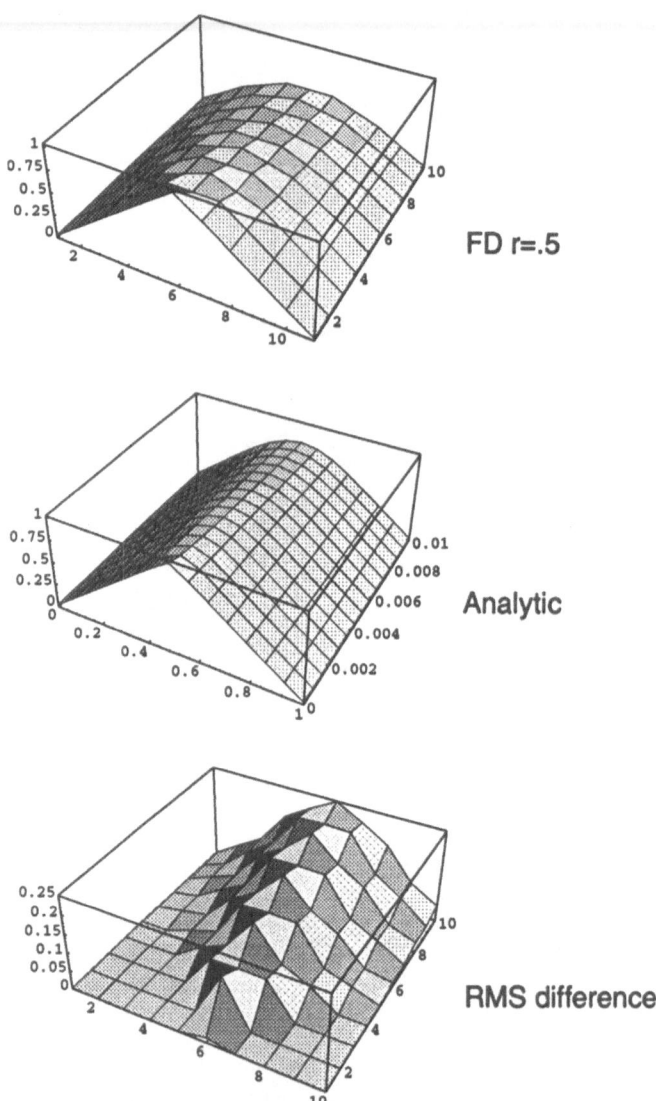

Fig. 12.2. Finite difference and analytic solution to the initial-boundary value problem for the heat equation described in the text. The discretization factor $r = .5$. The errors are substantial and growing with time.

A finite difference scheme will remain stable if the error e_j remains bounded as $j \to \infty$. This, in turn, requires that the eigenvalues of $\mathbf{A}$ be less than 1, since e_j results from the repeated application of the matrix to an arbitrary vector.[1]

Let us rewrite the tridiagonal matrix $\mathbf{A}$ as the identity matrix plus r times the tridiagonal matrix $\mathbf{T}_{n-1}$ where

$$
\mathbf{T}_{n-1} =
\begin{bmatrix}
-2 & 1 & 0 & 0 & \cdots \\
1 & -2 & 1 & 0 & \cdots \\
0 & 1 & -2 & 1 & \cdots \\
\vdots & & & & \ddots \\
0 & \cdots & 0 & 1 & -2
\end{bmatrix}.
$$

The matrix $\mathbf{T}_{n-1}$ has been well-studied and its eigenvalues and eigenvectors are known exactly ([42], p. 85ff). In particular its eigenvalues are $-4\sin^2\left(\frac{s\pi}{2N}\right)$. The more general result is that the eigenvalues of the $N \times N$ tridiagonal

$$
\begin{bmatrix}
a & b & 0 & 0 & \cdots \\
c & a & b & 0 & \cdots \\
0 & c & a & b & \cdots \\
\vdots & & & & \ddots \\
0 & \cdots & 0 & c & a
\end{bmatrix}
$$

are

$$
\lambda_s = a + 2\sqrt{bc}\cos\left(\frac{s\pi}{N+1}\right).
$$

This means that λ_s, the eigenvalues of $\mathbf{A}$, are given by

$$
\lambda_s = 1 + r\left\{-4\sin^2\left(\frac{s\pi}{2N}\right)\right\}.
$$

---

[1] Suppose $\mathbf{x}_s$ is the s-th eigenvector of $\mathbf{A}$ and λ_s is the s-th eigenvalue. Any vector, in particular the initial error vector, can be written as a summation of eigenvectors (assuming the matrix is full rank) with coefficients c_s:

$$
e_0 = \sum c_s \mathbf{x}_s.
$$

Therefore

$$
e_j = \sum c_s \lambda_s^j \mathbf{x}_s.
$$

So the question is, for what values of r is this expression less than or equal to one? It is not difficult to show that we must have $4r \sin^2 \left(\frac{s\pi}{2N} \right) \leq 2$, and hence if $r \leq 1/2$ then $\lambda_s \leq 1$.

This feature of the explicit time-integration scheme, that the stepsize must be strictly controlled in order to maintain stability, is common of all explicit schemes, whether for the parabolic heat equation or the hyperbolic wave equation. It is possible, however, to derive finite difference schemes which are unconditionally stable. I.e., stable for all finite values of r. It is important to keep in mind however, that just because a method might be stable for a large value of r, does not mean that it will be accurate! But, as we will now show, it is possible to derive finite difference methods where the constraints on r depend only on the desired accuracy of the scheme and not its stability.

12.1.2 Implicit Time-Integration Schemes

The method of Crank and Nicolson involves replacing the second derivative term in the heat equation with its finite difference representation at the j-th and $j+1$-th time steps. In other words, instead of Equation (12.11), we have:

$$\frac{u_{i,j+1} - u_{i,j}}{k} = \tag{12.13}$$

$$\frac{1}{2} \left[\frac{u_{i+1,j+1} - 2u_{i,j+1} + u_{i-1,j+1}}{h^2} + \frac{u_{i+1,j} - 2u_{i,j} + u_{i-1,j}}{h^2} \right]$$

This is an example of an implicit time-integration scheme. It is said to be implicit because the finite difference equations define the solution at any given stage implicitly in terms of the solution of a linear system of equations such as Equation (12.13).

Using the same notation as for the explicit case, we can write the Crank-Nicolson time-integration as

$$(2\mathbf{I} - r\mathbf{T}_{n-1}) \, \mathbf{u}_{j+1} = (2\mathbf{I} + r\mathbf{T}_{n-1}) \, \mathbf{u}_j.$$

Formally the solution of this set of equations is given by

$$\mathbf{u}_{j+1} = (2\mathbf{I} - r\mathbf{T}_{n-1})^{-1} \, (2\mathbf{I} + r\mathbf{T}_{n-1}) \, \mathbf{u}_j.$$

We don't need to calculate the inverse of the indicated matrix explicitly. For purposes of the stability analysis it suffices to observe that the eigenvalues of the matrix

$$\mathbf{A} \equiv (2\mathbf{I} - r\mathbf{T}_{n-1})^{-1}(2\mathbf{I} + r\mathbf{T}_{n-1})$$

will have modulus less than one if and only if

$$\frac{2 - 4r \sin^2 \left(\frac{s\pi}{2N}\right)}{2 + 4r \sin^2 \left(\frac{s\pi}{2N}\right)} \le 1. \tag{12.14}$$

And this is clearly true for any value of r whatsoever.

The fact that these eigenvalues have modulus less than 1 for all values of r implies *unconditional stability* of the Crank-Nicolson scheme. However, do not confuse this stability with accuracy. We may very well be able to choose a large value of r and maintain stability, but that does not mean the finite difference solution that we compute will accurately represent the solution of the PDE. Now, clearly δx and δt must be controlled by the smallest wavelengths present in the solution. So it stands to reason that if the solution of the PDE varies rapidly with space or time, then we may need to choose so small a value of the grid size that we might as well use an explicit method. The tradeoff here is that with Crank-Nicolson, we must solve a tridiagonal system of equations to advance the solution each time step.

But solving a tridiagonal, especially a symmetric one, can be done very efficiently. Algorithms for the efficient application of Gaussian elimination can be found in *Numerical Recipes* or the *Linpack Users' Guide*. The latter implementation is highly efficient. How can we judge this tradeoff? If we count the floating point operations (flop) implied by the explicit time-integration scheme Equation (12.11), we arrive at $5N$ floating point operations per time step, where N is the number of x gridpoints.[2] Now δx and hence N can be taken to be the same for both the explicit and the implicit methods. The difference will come in δt. The number of floating point operations required to perform Gaussian elimination on a symmetric tridiagonal is approximately $(13+4F)\frac{N-1}{2}$ where F is the number of floating point operations required to do a divide. This will vary from machine to machine but is likely to be in the range of 10-20. Now N will be large, so we can ignore the difference between $N-1$ and N. Therefore the ratio of work between the explicit and the implicit methods will be approximately

$$\frac{5N}{(13 + 4F)\frac{N}{2}} = \frac{10}{13 + 4F} \tag{12.15}$$

per time step. But the whole point is that we may be able to use a much larger time step for the implicit method. Suppose, for example, that we need to use an r of .1 for the explicit methods, but that we can use an r of 1 for the implicit approach. That means that we'll need ten times as many time steps for the former

---

[2] We count a multiply as one flop and an add as one flop. So the expression $ax + y$ implies 2 flops.

as the latter. Therefore the ratio of work between the two methods to integrate out to a fixed value of time is

$$\frac{100}{13 + 4F} \tag{12.16}$$

or more generally

$$\frac{\rho 10}{13 + 4F} \tag{12.17}$$

where ρ is the ratio of the number of explicit time steps to the number of implicit time steps.

Explicit/Implicit ratio of work

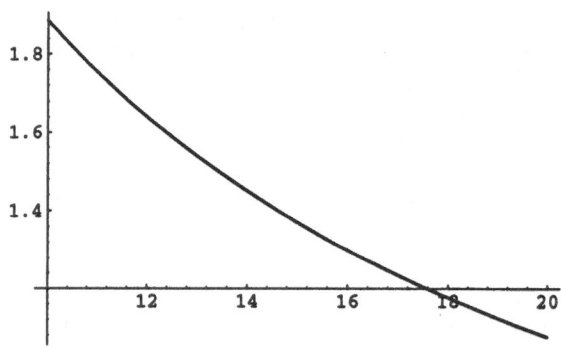

Floating point operations per divide

Fig. 12.3. Assuming $r = .1$ for the explicit method and $r = 1$ for the implicit, here is a plot showing the ratio of work between the two methods as a function of the number of floating point operations per divide on a particular computer. With a typical value of 15 or 16 floating point operations per divide, we can see that the work required for the two methods is very similar notwithstanding the fact that we need to use ten times as many time steps in the explicit case.

12.1.3 Weighted Average Schemes

A class of finite difference approximations that includes both the fully explicit and the Crank-Nicolson methods as special cases is based upon a weighted average of the spatial derivative terms

$$\frac{u_{i,j+1} - u_{i,j}}{k} = \tag{12.18}$$

$$\theta\left(\frac{u_{i+1,j+1} - 2u_{i,j+1} + u_{i-1,j+1}}{h^2}\right) + (1-\theta)\left(\frac{u_{i+1,j} - 2u_{i,j} + u_{i-1,j}}{h^2}\right).$$

If $\theta = 0$ we are left with the usual explicit scheme, while for $\theta = .5$ we have Crank-Nicolson. For $\theta = 1$ we have a so-called fully implicit method. For $\theta > .5$ the method is unconditionally stable, while for $0 \le \theta \le .5$, stability requires that

$$r \equiv \frac{\delta t}{(\delta x)^2} \le \frac{1}{2(1 - 2\theta)}.$$

12.1.4 Separation of Variables for Difference Equations

It is possible to give an analytic solution to the explicit finite difference approximation of the heat equation in the case that the solution to the PDE is periodic on the unit interval. Just as in the case of classical separation of variables, we assume a solution of the difference equation

$$u_{i,j+1} = r u_{i-1,j} - (1 - 2r)u_{i,j} + r u_{i+1,j} \tag{12.19}$$

of the form

$$u_{i,j} = f_i g_j$$

so that

$$\frac{g_{j+1}}{g_j} = \frac{r f_{i-1} + (1 - 2r)f_i + r f_{i+1}}{f_i}.$$

As in the standard approach to separation of variables, we observe at this point that the left side of this equation depends only on j while the right side depends only on i. Therefore, both terms must actually be constant.

$$\frac{g_{j+1}}{g_j} = \frac{r f_{i-1} + (1 - 2r)f_i + r f_{i+1}}{f_i} = C$$

where C is the separation constant. The solution to

$$g_{j+1} - C g_j = 0$$

is

$$g_j = AC^j.$$

The other equation

$$f_{i+1} + \frac{1 - 2r - C}{r} f_i + f_{i-1} = 0$$

presumably must have a periodic solution since the original PDE is periodic in space. Therefore a reasonable guess as to the solution of this equation would be

$$f_i = B\cos(i\theta) + D\sin(i\theta).$$

We won't go through all the details, but it can be shown using the known eigenvalues of the tridiagonal that

$$C = 1 - 4r\sin^2\left(\frac{s\pi}{2N}\right)$$

and, hence, a particular solution of the difference equation is

$$u_{i,j} = E\left(1 - 4r\sin^2\left(\frac{s\pi}{2N}\right)\right)^j \sin\left(\frac{si\pi}{N}\right).$$

We can generalize this to a particular solution for given initial data by taking the coefficients E_s to be the Fourier coefficients of the initial data. Then we have

$$u_{i,j} = \sum_{s=1}^{\infty} E_s \left(1 - 4r\sin^2\left(\frac{s\pi}{2N}\right)\right)^j \sin\left(\frac{si\pi}{N}\right).$$

12.2 Hyperbolic Equations

Now let us consider the first-order hyperbolic equation

$$U_t + cU_x = 0 \tag{12.20}$$

where, for the time being, c is a constant. Later we will generalize the discussion to allow for arbitrarily heterogeneous wave speed.

Along the line defined by $x - ct \equiv \xi = \text{constant}$, we have

$$\frac{dU}{dt} = \frac{d}{dt}[U(ct + \xi, t)] = cU_x + U_t \equiv 0. \tag{12.21}$$

Therefore U is constant along such lines. Different values of ξ give different such lines. As a result, the general solution of Equation (12.20) can be written

$$U(x,t) = f(x - ct) = f(\xi) \tag{12.22}$$

where the function f defines the initial data, since $U(x,0) = f(x)$.

The lines $x - ct$ define what are known as *characteristics*. The value of the solution U at an arbitrary point (x,t) depends only on the value of f at the single argument ξ. In other words, the *domain of dependence* of U on its initial values is given by the single point ξ. This is illustrated in Figure 12.4.

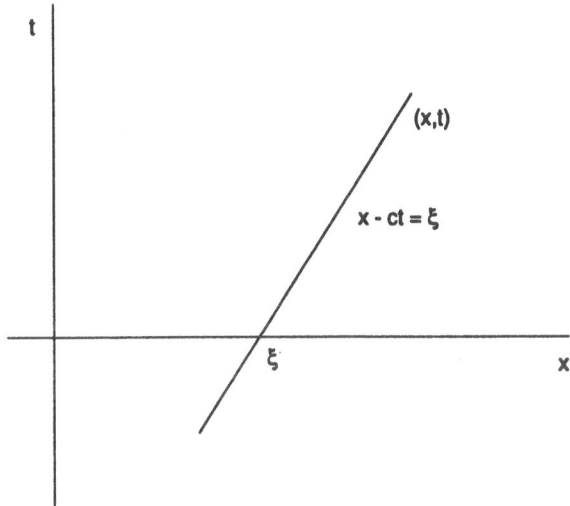

Fig. 12.4. Domain of dependence of the first-order hyperbolic equation.

Following John [31], we now show that a simple first-order forward-difference scheme of the type employed for the heat equation leads to an unstable algorithm when applied to a hyperbolic equation. This leads to the notion that the domain of dependence of a difference equation must be the same as that of the underlying PDE: this is the *Courant–Friedrichs–Lewy* condition.

Proceeding in the same fashion as for the heat equation, let us consider the following first-order finite-difference approximation of Equation (12.20)

$$u_{i,j+1} = u_{i,j} - rc[u_{i+1,j} - u_{i,j}] \tag{12.23}$$

where $r = k/h = \delta t / \delta x$. Defining the shift operator E via $Eu_{i,j} = u_{i+1,j}$ we have

$$u_{i,j} = ((1 + rc) - rcE)\, u_{i,j-1} \tag{12.24}$$

and hence by induction

$$u_{i,N} = ((1 + rc) - rcE)^N\, u_{i,0} \tag{12.25}$$

or

$$u(x,t) = u(x, Nk) = ((1 + rc) - rcE)^N\, f(x). \tag{12.26}$$

Using the binomial theorem we have

$$u(x,t) = \sum_{m=0}^{N} \binom{N}{m} (1 + rc)^m (-rcE)^{N-m} f(x) \tag{12.27}$$

and hence

$$u(x,t) = \sum_{m=0}^{N} \binom{N}{m} (1 + rc)^m (-rc)^{N-m} f(x + (N - m)h). \tag{12.28}$$

The first indication that something is wrong with this scheme comes from the observation that the domain of dependence of the difference equation appears to be

$$x, x + h, x + 2h, \cdots, x + Nh$$

whereas the domain of dependence of the PDE is $\xi = x - ct$, which in this case is $x - cNk$. Since this is not even in the set $x + ih$, it is clear that in the limit as $k \to 0$ the difference scheme cannot converge to the solution of the PDE. Further, the errors associated with some perturbed value ϵ of the initial data grow exponentially[3] according to

$$\epsilon \sum_{m=0}^{N} \binom{N}{m} (1 + rc)^m (-rc)^{N-m} = (1 + 2rc)^N \epsilon.$$

On the other hand, if we use a backward difference approximation in space then we have

$$u(x,t) = u(x, Nk) = ((1 - rc) + rcE^{-1})^N\, f(x) \tag{12.29}$$

---

[3] This is a worst-case scenario in which the error has been cleverly chosen with alternating sign to cancel out the alternating sign in the binomial result.

and

$$u(x,t) = \sum_{m=0}^{N} \binom{N}{m} (1-rc)^m (rc)^{N-m} f(x-(N-m)h). \qquad (12.30)$$

Now the domain of dependence of the difference equation is

$$x, x-h, x-2h, \cdots, x-Nh = x - t/r.$$

This set has the interval $[x - t/r, x]$ as its limit as the mesh size goes to zero, provided the ratio r is fixed. Therefore the point $\xi = x - ct$ will be in this interval if

$$rc \le 1. \qquad (12.31)$$

This is referred to variously as the Courant–Friedrichs–Lewy condition, the CFL condition, or simply the Courant condition.

Using the backward difference scheme, Equation (12.30), we see that the error growth is governed by

$$\epsilon \sum_{m=0}^{N} \binom{N}{m} (1-rc)^m (rc)^{N-m} = (1+rc+rc)^N \epsilon = \epsilon$$

which gives us confidence in the stability of the method, provided the CFL criterion is satisfied.

12.3 The Full Two-Way Wave Equation

By this point it should be clear how we will solve the full wave equation by finite differences. First we show a finite difference formula for the wave equation in two space dimensions which is second-order accurate in both space and time. Then we show a scheme which is fourth-order accurate in space and second in time. In both cases we take the grid spacing to be the same in the x and z directions: $\delta x = \delta z$.

$$
\begin{aligned}
u_{i,j,k+1} \;=\;\; & 2(1-2C_{i,j})u_{i,j,k} \\
+\;\; & C_{i,j}\left[u_{i+1,j,k}+u_{i-1,j,k}+u_{i,j+1,k}+u_{i,j-1,k}\right] \\
-\;\; & u_{i,j,k-1}
\end{aligned}
\tag{12.32}
$$

where

$$
C_{i,j} = (c_{i,j}dt/dx)^2,
$$

$c_{i,j}$ is the spatially varying wavespeed, and where the first two indices in u refer to space and the third refers to time. The CFL condition for this method is

$$
\delta t \le \frac{\delta x}{c_{\max}\sqrt{2}}
$$

where $c_{\max}$ is the maximum wavespeed in the model.

A scheme which is fourth-order accurate in space and second-order accurate in time is

$$
\begin{aligned}
u_{i,j,k+1} \;=\;\; & (2-5C_{i,j})u_{i,j,k} \\
+\;\; & C_{i,j}/12\left\{16\left[u_{i+1,j,k}+u_{i-1,j,k}+u_{i,j+1,k}+u_{i,j-1,k}\right]\right. \\
-\;\; & \left.[u_{i+2,j,k}+u_{i-2,j,k}+u_{i,j+2,k}+u_{i,j-2,k}]\right\} \\
-\;\; & u_{i,j,k-1}.
\end{aligned}
\tag{12.33}
$$

The CFL condition for this method is

$$
\delta t \le \frac{\delta x}{c_{\max}}\sqrt{3/8}
$$

For more details, see [2]. A straightforward extension of these explicit finite difference schemes to elastic wave propagation is contained in the paper by Kelly et al. [34]. This paper exerted a profound influence on the exploration geophysics community, appearing, as it did, at a time when a new generation of digital computers was making the systematic use of sophisticated finite difference modeling feasible.

Strictly speaking, the boxed finite difference equations above are derived assuming a homogeneous wavespeed. Nevertheless, they work well for heterogeneous models in some cases, especially if the heterogeneities are aligned along the computational grid. More general heterogeneous formulations require that we start with a wave equation which still has the gradients of the material properties. Although this generalization is conceptually straightforward, the resulting finite difference equations are a lot more complicated. For details, see [34]. In particular, this paper gives a finite difference scheme valid for P-SV wave propagation in arbitrarily heterogeneous 2D media. (It requires more than a full page just to write it down.)

12.3.1 Grid Dispersion

The question remains as to how fine the spatial grid should be in order to avoid numerical dispersion of the sort we first studied in the section on the 1D lattice (page 72 ff.). This measure of the grid size for dispersion purposes is given in terms of the number of gridpoints per wavelength contained in the source wavelet. But which wavelength, that associated with Nyquist, with the peak frequency, or some other? Alford *et al.* introduced the convention of defining the number of grid points

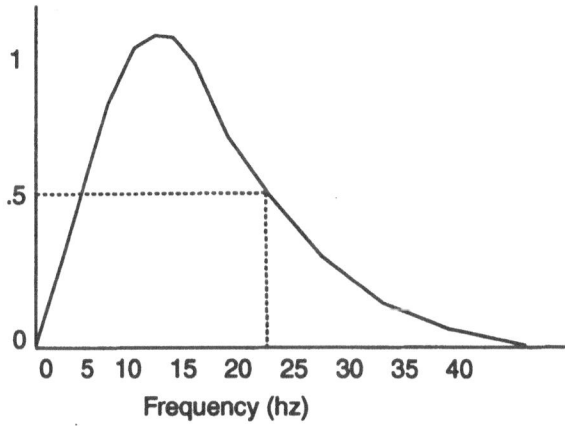

Fig. 12.5. Hypothetical normalized source power spectrum. The number of grid points per wavelength may be measured in terms of the wavelength at the upper half-power frequency.

per wavelength in terms of the wavelength at the upper half-power frequency. This is illustrated in Figure 12.5 for a hypothetical source wavelet. Measured in these terms, it is found that for best results at least 10-11 grid points per wavelength are required for the second-order accurate scheme shown in the first box above. Whereas, using the fourth-order scheme, this could be relaxed to around 5 grid points per wavelength. This means that using the fourth-order scheme reduces the number of grid points by a factor of 4, whereas the number of floating point operations per time step goes up by only 50%.

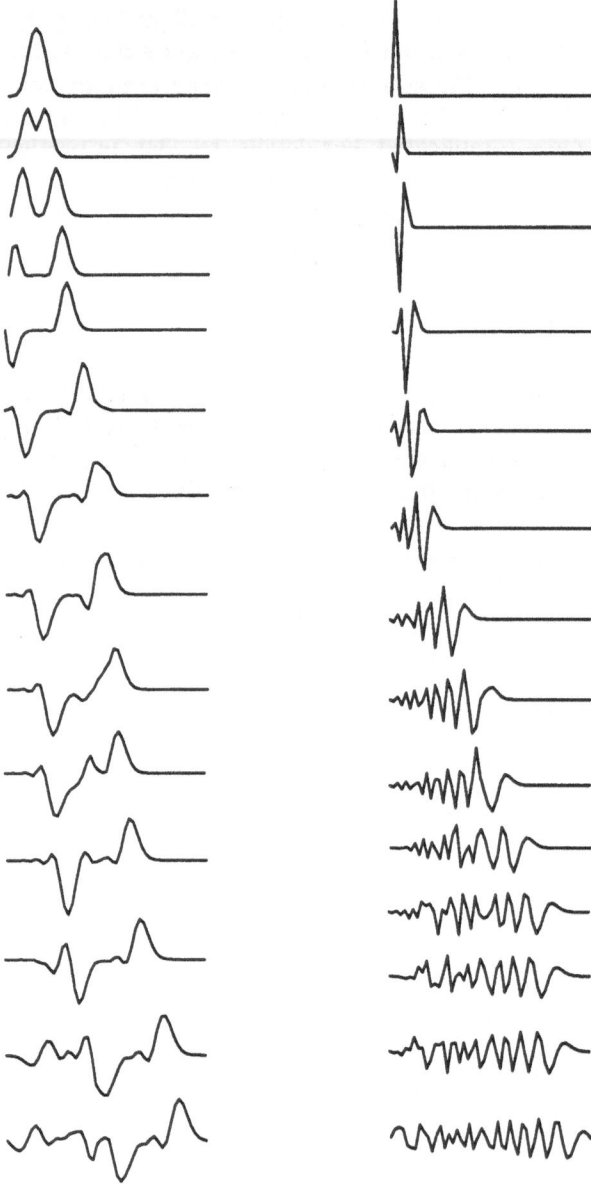

Fig. 12.6. Waves on a discrete string. The two columns of figures are snapshots in time of waves propagating in a 1D medium. The only difference between the two is the initial conditions, shown at the top of each column. On the right, we are trying to propagate a discrete delta function (one grid point in space and time). On the left, we are propagating a band-limited version of this. The medium is homogeneous except for an isolated reflecting layer in the middle. The dispersion seen in the right side simulation is explained by looking at the dispersion relation for discrete media. Waves whose wavelengths are comparable to the grid spacing sense the granularity of the medium.

12.4 Absorbing Boundary Conditions

We won't go into much detail on the problem of absorbing boundary conditions. The idea is quite simple. Whereas the real Earth is essentially unbounded, except at the free-surface, our computational grid needs to be as small as possible to save storage. If we didn't do anything special at the sides and bottom of the model, the wave propagating through this grid would think it was seeing a free surface when it reached one of these three sides. Clayton and Engquist [15] proposed replacing the full wave equation with its paraxial approximation near the non-free-surface edges of the model. Using a paraxial equation ensures one-way propagation at the edges of the model. Since the paraxial approximations they use are first order in the spatial extrapolation direction, only the nearest interior row of grid points is needed for each absorbing boundary.

Making paraxial approximations defines a preferred direction of propagation, typically along one of the spatial axes. This means that waves incident normally upon one of the absorbing sides of the model can be attenuated quite nicely. But wave incident at an angle will only be approximately attenuated. If one knew the direction of propagation of the waves which were to be attenuated, it would be possible to rotate coordinates in such a way that waves propagating in that direction were preferentially attenuated. Similarly, Clayton and Engquist show how a rotation can be used to take special account of the corners of the model. The detailed formulae for the Clayton-Engquist scheme are given in [15] and implemented in an example in the next section. The key idea, however, is that we use a standard explicit time integration scheme for all the interior points except for those immediately next to one of the sides we wish to be absorbing.

12.5 Finite Difference Code for the Wave Equation

In this appendix we give an extremely simple, but nonetheless functioning, code for solving the acoustic wave equation via finite differences. This code is for 2D and is second-order accurate in both space and time. It is written in Fortran 90 for use on the Connection Machine. It runs at around 3 Gigaflops on a full CM-200 and was written by Jacek Myzskowski of Thinking Machines Inc.

```
c      Fortran 90 implementation of a second order
c      (space-time) explicit finite difference scheme.
c      In this version the boundary conditions are
c      ignored.  The ony reason for using Fortran 90
c      in this case is that it runs fast on a
```

```
c        Connection Machine.

         parameter (f0=60.)
         parameter (nx=512,ny=512,nt=666,nrate=1,nsnap=4)
         real p1(nx,ny),p2(nx,ny),p3(nx,ny),c(nx,ny),pmax
         logical mask(nx,ny)
         pi = 4.*atan(1.)

           print*,'make a simple layered model'
           c(1:nx,1:ny/4) = 2.
           c(1:nx,ny/4+1:ny/2) = 3.
           c(1:nx,ny/2+1:3*ny/4) = 4.
           c(1:nx,3*ny/4+1:ny) = 5.
           print*,'finish model generation'
             cmin = minval(c)
             cmax = maxval(c)
           print*,cmin,cmax,'cmin,cmax'
c
c        11 grid points per wavelength, max frequency = 2*f0
c
         dx = cmin/(11.*2.*f0)
         dt = dx/(cmax*sqrt(2.))
         dtbydx=dt/dx
         c = c*dtbydx
         c = c*c

c        put a point source in the middle of the grid
         ix0 = nx/2
         iy0 = ny/2

c        Start time iterations

c
c        Make the ricker wavlet
c
         t0 = 1.5/f0
         w0 = 2*pi*f0
c
         mask = .false.
         mask(ix0,iy0) = .true.

c        begin time integration
         do 100 it=1,nt
c
```

```
c     Calculate the source amplitude
c
      t = (it - 1)*dt - t0
      arg1 = min(15.,(w0*t/2.)**2)
      arg2 = min(15.,(w0*(t + pi/w0)/2.)**2)
      arg3 = min(15.,(w0*(t - pi/w0)/2.)**2)
      amp = (1 - 2*arg1)*exp(-arg1)
      amp = amp - .5*(1 - 2*arg2)*exp(-arg2)
      amp = amp - .5*(1 - 2*arg3)*exp(-arg3)
c
c     Loop over space
c
      p3 =   (2-4*c)*p2
     &               + c*(cshift(p2,2,-1)+cshift(p2,2,1)
     &               + cshift(p2,1,-1)+cshift(p2,1,1) )
     &               - p1
      where(mask) p3 = p3 + amp

      p1=p2
      p2=p3

100   continue


      stop
      end
```

In a more sophisticated version, we define auxilliary arrays of weights in order to effect Clayton-Engquist absorbing boundary conditions on three of the four sides of the computational grid. So first we define the weights, then we put them into the finite difference stencil.

. . . . .


```
c
c     Precompute this to save run time
c     Inside weights
c
      p3=c*c
      c2 = 2-4*p3
      c1 = p3
```

```
            c3 = p3
            c5 = p3
            c4 = p3
            c6 = -1
            p3 = 0.

        if(itype.eq.2) then
c Top side weights
            mask = .false.
            mask(1,3:ny-2) = .true.
            where(mask)
              c1 = 0
              c2 = 0
              c3 = 0
              c4 = 0
              c5 = 0
              c6 = 0
            end where

c Bottom side weights
            mask = cshift(mask,1,1)
            where(mask)
              c2 = 1-c
              c1 = 0
              c3 = 0
              c4 = c
              c5 = 0
              c6 = 0
            end where

c Left side weights

            mask = .false.
            mask(3:nx-2,1) = .true.
            where(mask)
              c2 = 1-c
              c1 = 0
              c3 = c
              c5 = 0
              c4 = 0
              c6 = 0
            end where

c Right side weights
```

```
        mask = cshift(mask,2,1)
        where(mask)
          c2 = 1-c
          c1 = c
          c3 = 0
          c5 = 0
          c4 = 0
          c6 = 0
        end where

c Right lower corner weights
        mask = .false.
        mask(nx,ny) = .true.
        mask(nx-1,ny) = .true.
        mask(nx,ny-1) = .true.
        where(mask)
          c2 = 1-2*c
          c1 = c
          c3 = 0
          c4 = c
          c5 = 0
          c6 = 0
        end where

c Right upper corner weights
        mask(nx-1,ny) = .false.
        mask = cshift(mask,1,-1)
        mask(2,ny) = .true.
        where(mask)
          c2 = 1-2*c
          c1 = c
          c3 = 0
          c4 = 0
          c5 = c
          c6 = 0
        end where

c Left upper corner weights
        mask(1,ny-1) = .false.
        mask = cshift(mask,2,-1)
        mask(1,2) = .true.
        where(mask)
          c2 = 1-2*c
          c1 = 0
```

```
            c3 = c
            c4 = 0
            c5 = c
            c6 = 0
          end where

c Left lower corner weights
          mask(2,1) = .false.
          mask = cshift(mask,1,1)
          mask(nx-1,1) = .true.
          where(mask)
            c2 = 1-2*c
            c1 = 0
            c3 = c
            c4 = c
            c5 = 0
            c6 = 0
          end where

      end if

..... stuff deleted

c
c     Loop over space
c
          p3 =   c2*p2
     &           +c1*cshift(p2,2,-1)+c3*cshift(p2,2,1)
     &           +c4*cshift(p2,1,-1)+c5*cshift(p2,1,1)
     &           +c6*p1
          where(mask) p3 = p3 + amp

      p1=p2
      p2=p3


100   continue

....etc
```

12.6 Computer Exercises: VI

Figure 12.7 shows the first common source gather from a synthetic data set containing 50 such gathers.[4] Each gather has 100 traces whose receiver locations are fixed throughout the experiment. The source moves across the model from left to right. The receiver spacing is 100. The source spacing is 200. The traces were recorded at 4 ms sampling, 1001 samples per trace. Your job is to discover the model and produce a migrated image using your now-fancy Kirchhoff migration codes.

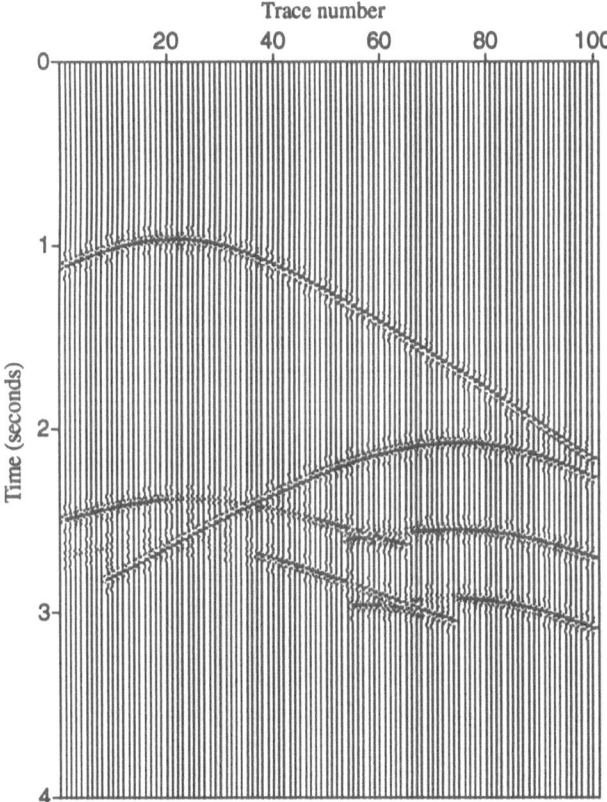

Fig. 12.7. The first of 50 common source gathers that comprise the final computer exercise. Figure out the model and produce a migrated image. Drilling is optional.

---

[4] These data are available via ftp from hilbert.mines.colorado.edu in the directory pub/data.

12.7 Computer Exercises: VI – Solution

Figure 12.8 shows the model that was used to produce the data in Figure 12.7. In the next two figures, we see the results obtained by Wences Gouveia and Tariq Alkhalifah. The recording aperature was limited strictly to the 1 km shown, so no reflections from the outside of the structure were recorded. Because of the ray bending and the dip of the syncline, it is very difficult to get a good image of the sides of the syncline and the dipping reflectors below the syncline. These are both extremely good images under the circumstances. Further, no information was given on the velocity model; all this had to be deduced from the data.

Here is the processing sequence that Gouveia applied to achieve the result shown in Figure 12.9.

1. NMO correction with a constant velocity (the velocity of the first layer, estimated by a sequence of Stolt migrations), as a pre-processing for DMO.

2. Constant velocity DMO in the frequency domain, to remove the effect of dip on the stacking velocities.

3. Inverse NMO with the velocity used in step 1.

4. Velocity analysis, via the standard semblance methods.

5. Conversion of the stacking velocities to interval velocities using the Dix equation as a first approximation.

6. Post stack depth migration (using modified Kirchhoff for handling lateral velocity variations) with the velocities estimated in the previous item. Refinements on this velocity profile was accomplished by repeating the post stack migration for slightly different velocity models.

And here is Alkhalifah's account of his efforts to achieve the results shown in Figure 12.10.

Because the synthetic data containing only three layers seemed at first to be reasonably easy, I went through a normal sequence. This sequence is based on applying constant velocity NMO, constant velocity DMO (this step is independent of velocity), inverse NMO, velocity analysis, NMO with velocities obtained from velocity analysis, stack, and finally Kirchhoff post-stack migration. Caustics were apparent in the data and only with a multi-arrival traveltime computed migration, such as a Kirchhoff migration based on ray tracing, that we can handle the caustics. The model also had large lateral velocity variations in which the velocity analysis and NMO parts of the above mentioned sequence can not handle. Such limitations caused clear amplitude degradations on the data especially for the deeper events (these events are mostly effected by the lateral velocity variations). The velocities used in the migration were obtained from velocity analysis and a lot of guessing. Also running the migration with different velocities helped in figuring out which velocities gave the best final image. After all is said and done, the model did not seem easy at all. It might have been straightforward when one uses a prestack migration with a migration velocity analysis, however with the tools I used it was basically a nightmare.

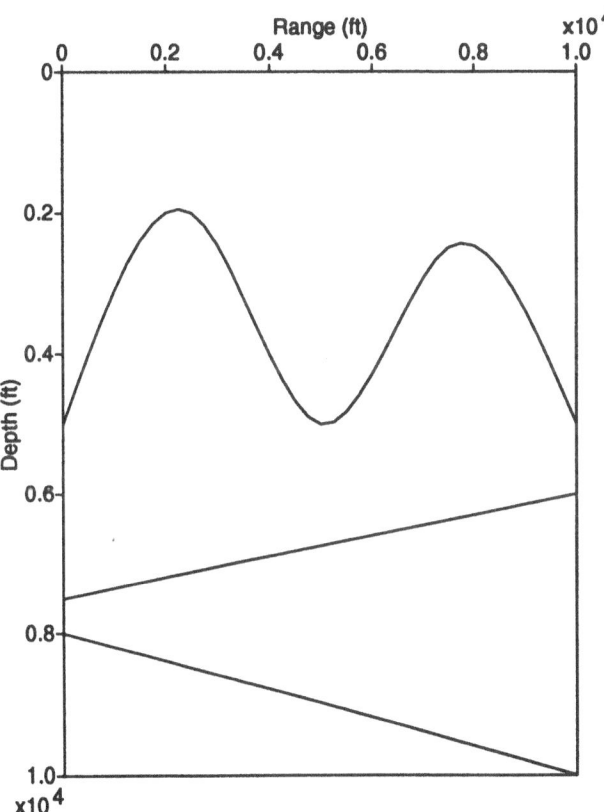

Fig. 12.8. The model that was used to produce the data shown in the previous figure.

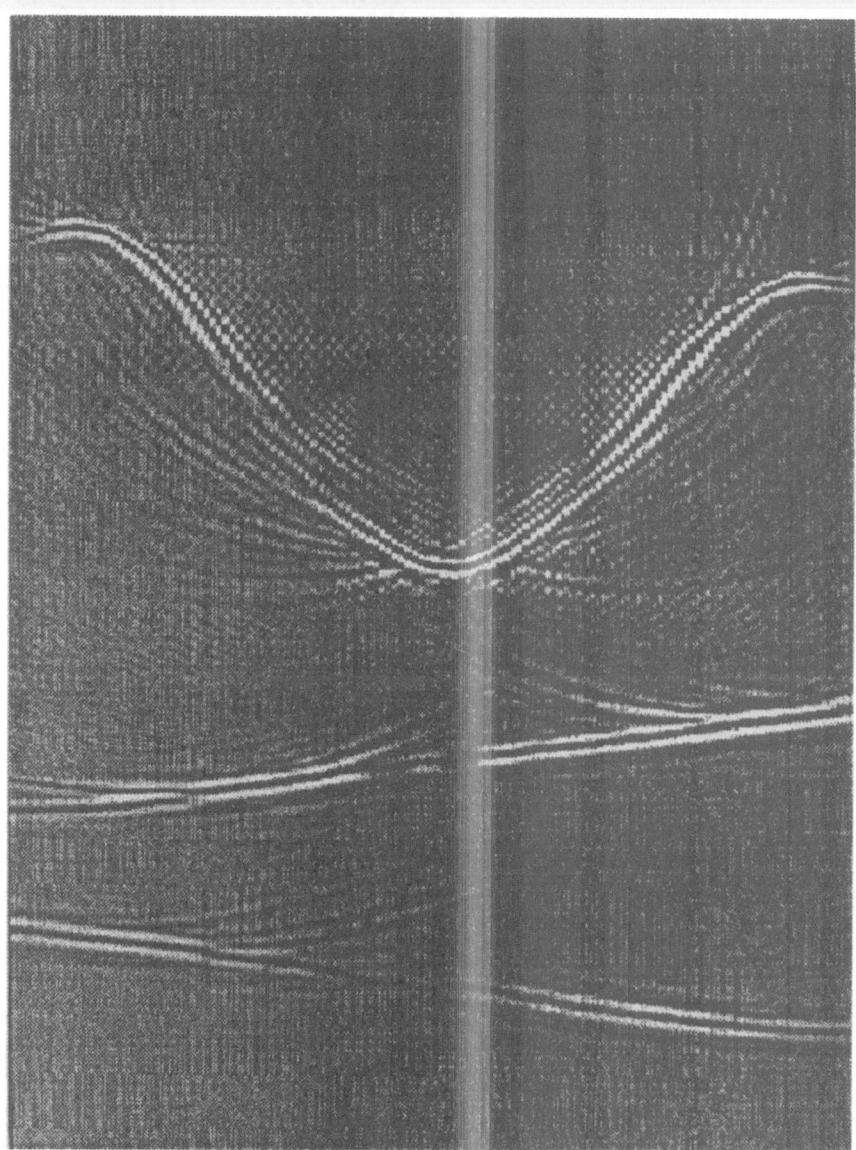

Fig. 12.9. Wences Gouveia's migrated depth section.

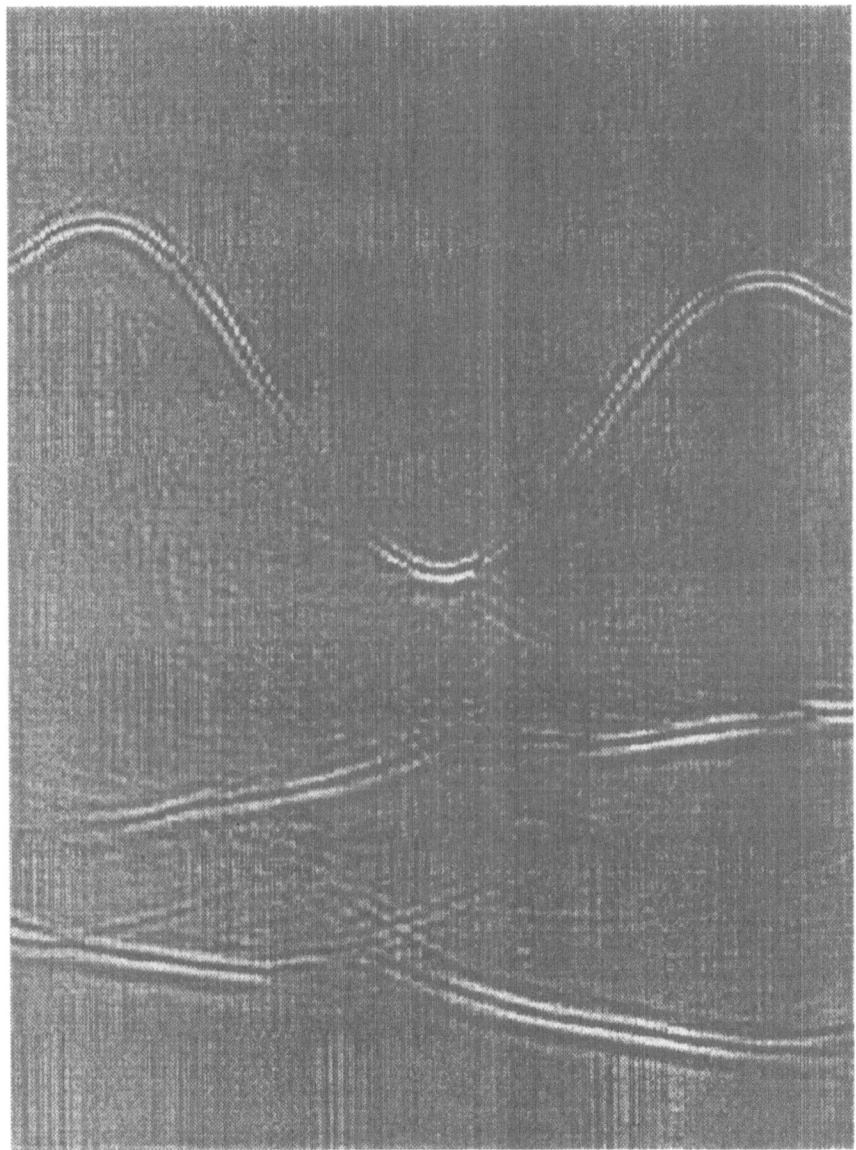

Fig. 12.10. Tariq Alkhalifah's migrated depth section.

A. SU User's Manual

Jack K. Cohen & John W. Stockwell
Center for Wave Phenomena
Colorado School of Mines
Golden, Colorado
USA
jkc@dix.mines.colorado.edu
john@dix.mines.colorado.edu

A.1 About SU

Seismic Unix (SU) is a self-contained software environment for seismic research and data processing used by exploration geophysicists, earthquake seismologists, environmental engineers, software developers and others. It is used by scientific staff in both small companies and major oil and gas companies, and by academics and government researchers.

The SU package is *free software*, distributed quarterly with the full source code, so that users can alter and extend its capabilities. The package permits the exchange of data according to the industry protocol (SEG-Y). It provides a standard environment for the testing of new processing algorithms. It is easy to use because it does not require learning a special language—its application uses only the standard facilities afforded by the UNIX operating system. Once UNIX shell-redirecting and pipes are mastered, there is no further artificial language to learn. The seismic commands and options can be used as readily as other UNIX commands. In particular, the user can write ordinary UNIX shell scripts to combine frequent command combinations into meta-commands (i.e., processing flows). These scripts can be thought of as "job files."

The seismic processing programs in the package assume that the data are written in SEG-Y format with each trace preceded by an appropriate header. This allows the data information to be read by each program in the processing stream in a consistent manner. The package includes facilities for converting data in several other formats to the SEG-Y format.

The SU user community accesses the software over the Internet using the commonly available "anonymous ftp" facility. In this way, users obtain the software with no constraints on its use. During installation, the user is given the option of sending an electronic mail request to add them to our user group. Members of the user group receive announcements when updates of the package become available. The SU community is now worldwide, spanning six continents, more than 26 countries and over 290 known installations on a variety of hardware platforms ranging from mainframes to workstations and PC's.

Parts of SU originated from software developed by students working with the Stanford Exploration Project at Stanford University. The present package was developed and is maintained at the Center for Wave Phenomena (CWP) at the Colorado School of Mines. CWP is an interdisciplinary (geophysics, mathematics) research and educational program in seismic exploration, supported by 23 companies in the oil and gas industry.

A.2 How to Get a Copy of SU

The SU package contains seismic processing programs along with libraries of scientific routines, graphics routines and routines supporting the SU coding conventions. The package is available by anonymous ftp at the site

    hilbert.mines.colorado.edu (138.67.12.63).

The directory path is pub/cwpcodes. Take the files:

1. README_BEFORE_UNTARRING

2. untar_me_first.xx.tar.Z

3. cwp.su.all.xx.tar.Z

Here the xx denotes the number of the current release. An incremental update is also available for updating the previous release yy to the current release xx. Take the files:

1. README_BEFORE_UNTARRING

2. README_UPDATE

3. untar_me_first.xx.tar.Z

4. update.yy.to.xx.tar.Z

5. update.list

For readers who are not familiar with anonymous ftp, an annotated transaction listing follows in section A.2.1.

A.2.1 Obtaining Files by Anonymous ftp

Type:
% ftp 138.67.12.63	—	138.67.12.63 is our ftp site
username: anonymous	—	username is "anonymous"
password: yourname@your.machine	—	type anything here
ftp>	—	this is the prompt you see when you are in ftp

You are now logged in via ftp to the CWP anonymous ftp site. You may type:

ftp> ls	—	to see the contents of the directories
ftp> cd dirname	—	to change directories to "dirname"
ftp> binary	—	set "binary mode" for transferring files You must do this before you transfer any binary file. This includes all files with the form some_name.tar.Z extension.
ftp> get filename	—	to transfer "filename" from our site to your machine
ftp> mget pattern*	—	to transfer all files with names of the "pattern*"
For example:		
ftp> mget *.tar.Z	—	will transfer all files with the form of name.tar.Z to your machine. You will be asked whether you really want each file of this name pattern transferred, before ftp actually does it.
ftp> bye	—	to exit from ftp

A.2.2 Requirements for Installing the Package

The only requirements for installing the package are:

1. A machine running the UNIX operating system.

2. Ten megabytes of disk space for the source and compiled binary.

The package has been successfully installed on:

- IBM RS6000

- SUN SPARC STATIONS

- HP 9000 series machines

- HP Apollo

- NeXT

- Convex

- DEC

- Silicon Graphics

- PC's running LINUX, PRIME TIME, ESIX, and NeXTSTEP 486

There are README files in the distribution with special notes about some of these platforms. We depend on the SU user community to alert us to installation problems, so if you run into difficulties, please let us know.

The distribution contains a series of files that detail the installation process. Read them in the following order:

```
LICENSE --- legal statement
README_BEFORE_UNTARRING --- initial information
README_FIRST --- general information
README_TO_INSTALL --- installation instructions
README_X --- X-windows install information
Portability/README_*    --- portability information
README_GETTING_STARTED --- how to begin using the codes
```

A.2.3 A Quick Test

Once you have completed the installation, here is a quick test you can make to see if you have a functioning seismic system. For an X-windows machine, the "pipeline"

```
suplane | suxwigb &
```

should produce the graphic shown in Figure A.1. If you have a PostScript printer, then you should get a hard copy version with the pipeline

```
suplane | supswigb | lpr
```

If you have display PostScript, then to get a screen display, replace the `lpr` command in the pipeline by the command that opens a PostScript file.

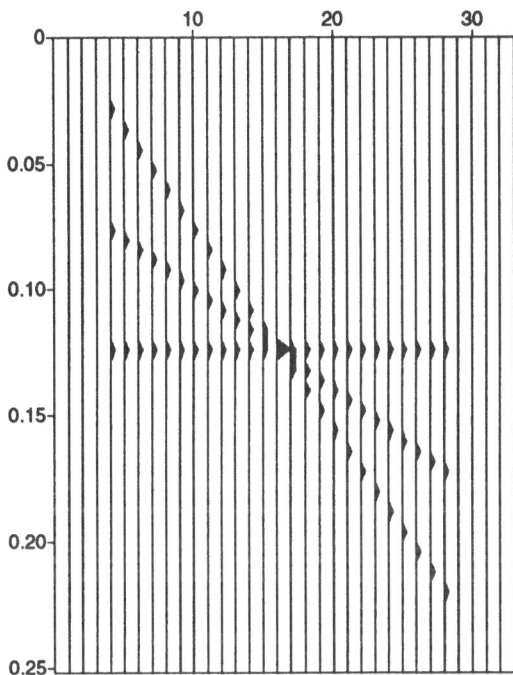

Fig. A.1. Output of the suplane pipeline.

A.3 Help Facilities

A.3.1 Self-Documentation (Selfdoc)

Virtually all SU programs give information about themselves when you type the name of the program without any arguments. For example,

```
% sustack
```

```
SUSTACK - stack adjacent traces having the same key header word

sustack <input >output key=cdp normpow=1.0 verbose=0
```

Required parameters:
        none

Optional parameters:
        key=cdp   header key word to stack on
        normpow=1.0   each sample is divided by the
                         normpow'th number of non-zero values
                         stacked (normpow=0 selects no division)
        verbose=0   verbose = 1 echos information

Note: The offset field is set to zero on the output traces.
        Sushw can be used afterwards if this is not acceptable.

A.3.2 Suhelp

suhelp lists the names of the programs in the distribution. The first "page" of
the output looks something like this:

```
% suhelp


CWP PROGRAMS: (no self-documentation)
ctrlstrip      fcat           maxints        t
downfort       isatty         pause          upfort

PAR PROGRAMS: (programs with self-documentation)
a2b            kaperture      resamp         transp         vtlvz
b2a            makevel        smooth2        unif2          wkbj
farith         mkparfile      smoothint2     unisam
ftnstrip       prplot         subset         unisam2
h2b            recast         swapbytes      velconv


press return key to continue
```

A.3.3 Suname

suname is a program that lists the names and a one-line description of all the
programs and libraries. Here is the first "page" of the output:

```
% suname
```

```
-----   CWP Free Programs -----

Mains:

In CWPROOT/cwp/main:
* CTRLSTRIP - Strip non-graphic characters
* DOWNFORT - change Fortran programs to lower case, preserving
             strings
* FCAT - fast cat with 1 read per file
* ISATTY - pass on return from isatty(2)
* MAXINTS - Compute maximum and minimum sizes for integer types
* PAUSE - prompt and wait for user signal to continue
* T - time and date for non-military types
* UPFORT - change Fortran programs to upper case, preserving
           strings

In CWPROOT/par/main:
A2B - convert ascii floats to binary
B2A - convert binary floats to ascii
FARITH - File ARITHmetic -- perform simple arithmetic with
         binary files
FTNSTRIP - convert a file of floats plus record delimiters
           created
H2B - convert 8 bit hexidecimal floats to binary
KAPERTURE - generate the k domain of a line scatterer for a
            seismic array
MAKEVEL - MAKE a VELocity function v(x,y,z)
MKPARFILE - convert ascii to par file format
--More--(5%)
```

Note that the directory paths are shown as an aid to the SU programmer.

A.3.4 Sudoc

sudoc is a utility that lists the self-documentation of all programs in the package. Subroutines and the few mains and shell scripts that do not show a selfdoc will have an sudoc entry.

For example,

```
% sudoc fgettr

In CWPROOT/src/su/lib:
FGETTR - Routines to get an SU trace from a file

fgettr          get a segy trace from a file by file pointer
gettr           macro using fgettr to get a trace from stdin

Function Prototype:
int fgettr(FILE *fp, segy *tp);
```

Returns:
int: number of bytes read on current trace (0 after last trace)

The function gettr(x) is a macro defined in su.h
define gettr(x) fgettr(stdin, (x))

Usage example:
```
      segy tr;
      ...
      while (gettr(&tr)) {
              tr.offset = abs(tr.offset);
              puttr(&tr);
      }
      ...
```

Authors: SEP: Einar Kjartansson, Stew Levin
CWP: Shuki Ronen, Jack Cohen

A.3.5 Sufind

sufind is a program that searches the self-documentations for a given string. For example,

```
% sufind fft

 FFTLAB - Motif-X based graphical 1D Fourier Transform

 Usage:  fftlab


HANKEL - Functions to compute discrete Hankel transforms

hankelalloc  allocate and return pointer to a Hankel transformer
hankelfree   free a Hankel transformer

PFAFFT - Functions to perform Prime Factor (PFA) FFT's, in place

npfa        return valid n for complex-to-complex PFA
npfar       return valid n for real-to-complex/complex-to-real PFA

 SUAMP - output amp, phase, real or imag trace from
        (frequency, x) domain data

 suamp <stdin >stdout mode=amp

 SUFFT - fft real time traces to complex frequency traces

 suftt <stdin >sdout sign=1
```

```
SUFRAC -- take fractional time derivative or integral of
              data, plus a phase shift.  Input is TIME DOMAIN data.

sufrac power= [optional parameters] <indata >outdata

SUIFFT - fft complex frequency traces to real time traces

suiftt <stdin >sdout sign=-1


SUMIGPS - MIGration by Phase Shift with turning rays

sumigps <stdin >stdout [optional parms]


SUMIGTK - MIGration via T-K domain method for common-midpoint
              stacked data

sumigtk <stdin >stdout dxcdp= [optional parms]


SURADON - forward generalized Radon transform from (x,t) ->
              (p,tau) space.

suradon <stdin >stdout [Optional Parameters]



For more information type: "program_name <CR>"
```

The final line of this output ends with a symbol meant to indicate that the user is to type a carriage return.[1]

A.3.6 Sukeyword

SU programs that manipulate the trace headers use specific names called "keywords" to identify the header fields. The sukeyword program enables the user to list the definition file for the keywords. For example,

```
% sukeyword fldr

...skipping
        int tracr;      /* trace sequence number within reel */

        int fldr;       /* field record number */
```

---

[1] The phrase "carriage return" refers to an older technology, the typewriter. Ask your parents for further details.

```
        int tracf;      /* trace number within field record */

        int ep;         /* energy source point number */

        int cdp;        /* CDP ensemble number */

        int cdpt;       /* trace number within CDP ensemble */

        short trid;     /* trace identification code:
                        1 = seismic data
                        2 = dead
                        3 = dummy
                        4 = time break
                        5 = uphole
                        6 = sweep
                        7 = timing
                        8 = water break
                        9---, N = optional use (N = 32,767)
  --More--(13%)
```

A.3.7 Other Help Mechanisms

- gendocs is a program that creates the LaTeX document, selfdocs.tex, that contains a complete set of all the self-documentations in the distribution (over 300 pages!). A PostScript version of this document is available in our anonymous ftp site
(pub/cwpcodes/documentation.xx.tar.Z).

Here, xx denotes the number of the current release.

- The demos directory contains a number of tutorial shell scripts. Its subdirectories contain README files that give detailed information. Assuming that you start in the demos directory, here is a roadmap to get you started:

The Sorting_Traces Tutorial is an interactive script that reinforces some of the basic UNIX and SU lore discussed in this document. The interactivity is limited to allowing you to set the pace. Such tutorials quickly get annoying, but we felt that one such was needed. The Sorting_Traces Demo is the format used in the rest of the demos directory—it just pops up a bunch of windows and assumes you'll read the script if you want to see how a particular window is evoked.

The next step is to activate the Selecting_Traces Demo. Then proceed to the Deconvolution Demo and the NMO Demo. Beyond that, visit the Demo directories that interest you. The demos directory tree is still under active development—please let us know if the demos are helpful and how they can be improved.

- The essence of SU usage is the construction of shell programs to carry out coordinated data processing. The su/examples directory contains a number of such programs. By the way, the terms "shell scripts," "shell programs," "shell files," and "shells," are used interchangeably in the UNIX literature.

– You should not hesitate to look at the source code itself. Please let us know if you discover any inconsistencies between the source and our documentation of it. We also welcome suggestions for improving the comments and style of our codes.

A.4 Using SU

A.4.1 SU and UNIX

You need not learn a special seismic language to use SU. If you know how to use UNIX shell-redirecting and pipes, you are ready to start using SU—the seismic commands and options can be used just as you would use the built-in UNIX commands. In particular, you can write ordinary UNIX shell scripts to combine frequent command combinations into meta-commands (i.e., processing flows). These scripts can be thought of as "job files."

Table A.1. UNIX Symbols

process1 < file1	process1 takes input from file1
process2 > file2	process2 writes on (new) file2
process3 >> file3	process3 appends to file3
process4 | process5	output of process4 is input to process5
process6 << text	take input from following lines

So let's begin with a capsule review of the basic UNIX operators as summarized in Table A.4.1. The symbols <, >, and >> are known as "redirection operators," since they redirect input and output into or out of the command (i.e., process). The symbol | is called a "pipe," since we can picture data flowing from one process to another through the "pipe." Here is a simple SU "pipeline" with input "indata" and output "outdata":

```
sufilter f=4,8,42,54 <indata |
sugain tpow=2.0 >outdata
```

This example shows a band-limiting operation being "piped" into a gaining operation. The input data set indata is directed into the program sufilter with the < operator, and similarly, the output data set outdata receives the data because of the > operator. The output of sufilter is connected to the input of sugain by use of the | operator.

The strings with the = signs illustrate how parameters are passed to SU programs. The program **sugain** receives the assigned value 2.0 to its parameter tpow, while the program **sufilter** receives the assigned four component *vector* to its parameter f. To find out what the valid parameters are for a given program, we use the self-doc facility.

By the way, space around the UNIX redirection and pipe symbols is optional—the example shows one popular style. On the other hand, spaces around the = operator are *not* permitted.

The first four symbols in Table A.4.1 are the basic grammar of UNIX ; the final << entry is the symbol for the less commonly used "here document" redirection. Despite its rarity in interactive use, SU shell programs are significantly enhanced by appropriate use of the << operator—we will illustrate this below.

Many built-in UNIX commands do not have a self-documentation facility like SU'—instead, most do have "man" pages. For example,

```
% man cat

CAT(1)                    UNIX Programmer's Manual                    CAT(1)



NAME
     cat - catenate and print

SYNOPSIS
     cat [ -u ] [ -n ] [ -s ] [ -v ] file ...

DESCRIPTION
     Cat reads each file in sequence and displays it on the stan-
     dard output.  Thus

               cat file

     displays the file on the standard output, and

               cat file1 file2 >file3
--More--
```

You need to know a bit more UNIX lore to use SU efficiently—we'll introduce these tricks of the trade in the context of the examples discussed later in this section.

A.4.2 Exploring SU

This section is a simulated example of an interactive session with SU.

Looking for DMO programs. Later we will discuss the construction of SU processing streams, our present purpose is just to illustrate how to look around for the raw materials for such streams. Let's by using sufind to see if there are any DMO programs:

```
% sufind dmo

  SUDMOFK - DMO via F-K domain (log-stretch) method for
            common-offset gathers

  sudmofk <stdin >stdout cdpmin= cdpmax= dxcdp= noffmix= [...]


  SUDMOTX - DMO via T-X domain (Kirchhoff) method for common-offset
            gathers

  sudmotx <stdin >stdout cdpmin= cdpmax= dxcdp= noffmix=
          [optional parms]


  SUFDMOD2 - Finite-Difference MODeling (2nd order) for acoustic
             wave equation

  sufdmod2 <vfile >wfile nx= nz= tmax= xs= zs= [optional parameters]


  SUSTOLT - Stolt migration for stacked data or common-offset
            gathers

  sustolt <stdin >stdout cdpmin= cdpmax= dxcdp= noffmix= [...]
```

The last two "hits" are spurious, but we see that two DMO programs have been found.

Getting information about SU programs. Use the self-doc facility to get more information about sudmofk:

```
% sudmofk

  SUDMOFK - DMO via F-K domain (log-stretch) method for
            common-offset gathers

  sudmofk <stdin >stdout cdpmin= cdpmax= dxcdp= noffmix= [...]

  Required Parameters:
  cdpmin        minimum cdp (integer number) for which to apply DMO
  cdpmax        maximum cdp (integer number) for which to apply DMO
  dxcdp         distance between adjacent cdp bins (m)
  noffmix       number of offsets to mix (see notes)

  Optional Parameters:
  tdmo=0.0      times corresponding to rms velocities in vdmo (s)
```

```
vdmo=1500.0   rms velocities corresponding to times in tdmo (m/s)
sdmo=1.0      DMO stretch factor; try 0.6 for typical v(z)
fmax=0.5/dt   maximum frequency in input traces (Hz)
verbose=0     =1 for diagnostic print
```

Notes:
Input traces should be sorted into common-offset gathers.
One common- offset gather ends and another begins when
the offset field of the trace headers changes.

The cdp field of the input trace headers must be the cdp
bin NUMBER, NOT the cdp location expressed in units of
meters or feet.

The number of offsets to mix (noffmix) should typically
equal the ratio of the shotpoint spacing to the cdp spacing.
This choice ensures that every cdp will be represented in
each offset mix. Traces in each mix will contribute through
DMO to other traces in adjacent cdps within that mix.

The tdmo and vdmo arrays specify a velocity function of
time that is used to implement a first-order correction
for depth-variable velocity. The times in tdmo must be
monotonically increasing.

For each offset, the minimum time at which a non-zero
sample exists is used to determine a mute time. Output
samples for times earlier than this mute time will be
zeroed. Computation time may be significantly reduced
if the input traces are zeroed (muted) for early times
at large offsets.

Trace header fields accessed: ns, dt, delrt, offset, cdp.

Viewing header field definitions. Note that the end of the last program description referred to "header fields"; these names are *not* standard and, as mentioned earlier, you can get definitions by using sukeyword. For example,

```
% sukeyword delrt
```

```
...skipping
                        may be positive or negative */

        short delrt;    /* delay recording time, time in ms between
                           initiation time of energy source and time
                           when recording of data samples begins
                           (for deep water work if recording does not
                           start at zero time) */

        short muts;     /* mute time--start */

        short mute;     /* mute time--end */
```

```
    unsigned short ns;    /* number of samples in this trace */

    unsigned short dt;    /* sample interval; in micro-seconds */

    short gain;    /* gain type of field instruments code:
                        1 = fixed
                        2 = binary
                        3 = floating point
                        4 ---- N = optional use */
```

--More--(53%)

Viewing Program Names. SU program names are often obscure (we aren't proud of this). Here's how to get help with remembering the exact name of a program when you recall a fragment of the name:

```
% sufind -n head

  SUADDHEAD - put headers on bare traces and set the tracl and
              ns fields
  UPDATEHEAD - update ../doc/Headers/Headers.all

For more information type: "program_name <CR>"
```

Recall also that suhelp and suname give comprehensive listings of the SU programs.

Note that we used the -n option of the sufind command. Using the self-doc facility, we can get the full story:

```
% sufind

sufind - get info from self-docs about SU programs
Usage: sufind [-v -n] string
"sufind string" gives brief synopses
"sufind -v string" verbose hunt for relevant items
"sufind -n name_fragment" searches for command name
```


A.4.3 Understanding and Using SU Shell Programs

The essence of good SU usage is constructing (or cloning!) UNIX shell programs to create and record processing flows. In this section, we give some annotated examples to get you started.

A simple SU processing flow example. Most SU programs read from standard input and write to standard output. Therefore, one can build complex processing flows by simply connecting SU programs with UNIX pipes. Most flows will end with one of the SU plotting programs. Because typical processing flows are lengthy and involve many parameter settings, it is convenient to put the SU commands in a shell file.

Remark: All the UNIX shells, Bourne (sh), Cshell (csh), Korn (ksh), ..., include a programming language. In this document, we exclusively use the Bourne shell programming language.

Our first example is a simple shell program called Plot. The numbers in square brackets at the end of the lines in the following listing are not part of the shell program—we added them as keys to the discussion that follows the listing.

```
#! /bin/sh                                           [1]
# Plot:   Plot a range of cmp gathers
# Author: Jane Doe
# Usage:  Plot cdpmin cdpmax

data=$HOME/data/cmgs                                 [2]

# Plot the cmp gather.
suwind <$data key=cdp min=$1 max=$2 |                [3]
sugain tpow=2 gpow=.5 |
suximage f2=0 d2=1 \                                 [4]
        label1="Time (sec)" label2="Trace number" \
        title="CMP Gathers $1 to $2" \
        perc=99 grid1=solid &                        [5]
```

Discussion of numbered lines:

1. The symbol # is the comment symbol—anything on the remainder of the line is not executed by the UNIX shell. The combination #! is an exception to this rule: the shell uses the file name following this symbol as a path to the program that is to execute the remainder of the shell program.

2. The author apparently intends that the shell be edited if it is necessary to change the data set—she made this easier to do by introducing the shell variable data and assigning to it the full pathname of the data file. The assigned value of this parameter is accessed as $data within the shell program. The parameter $HOME appearing as the first component of the file path name is a UNIX maintained environment variable containing the path of the user's home directory. In general, there is no need for the data to be located in the user's home directory, but the user would need "read permission" on the data file for the shell program to succeed.

WARNING! Spaces are significant to the UNIX shell—it uses them to parse command lines. So despite all we've learned about making code easy to read, do *not* put spaces next to the = symbol.

3. The main pipeline of this shell code selects a certain set of cmp gathers with suwind, gains this subset with sugain and pipes the result into the plotting program suximage. As indicated in the Usage comment, the cmp range is specified by command line arguments. Within the shell program, these arguments are referenced as $1, $2 (i.e., first argument, second argument).

4. The lines within the suximage command are continued by the backslash escape character.

WARNING! The line continuation backslash must be the *final* character on the line—an invisible space or tab following the backslash is one of the most common and frustrating bugs in UNIX shell programming.

5. The final & in the shell program puts the plot window into "background" so we can continue working in our main window. This is the X-Windows usage—the & should *not* be used with the analogous PostScript plotting programs (e.g., supsimage). For example, with supsimage in place of suximage, the & might be replaced by | lpr.

The SU plotting programs are special—their self-doc doesn't show all the parameters accepted. For example, most of the parameters accepted by suximage are actually specified in the self-documentation for the generic CWP plotting program ximage. This apparent flaw in the self-documentation is actually a side effect of a key SU design decision. The SU graphics programs call on the generic plotting programs to do the actual plotting. The alternative design was to have tuned graphics programs for various seismic applications. Our design choice keeps things simple, but it implies a basic limitation in SU's graphical capabilities.

The plotting programs are the vehicle for presenting your results. Therefore you should take the time to carefully look through the self-docs for *both* the "SU jacket" programs (suximage, suxwigb, ...) and the generic plotting programs (ximage, xwigb, ...).

Executing shell programs. The simplest way to execute a UNIX shell program is to give it "execute permission." For example, to make our above Plot shell program executable:

```
chmod +x Plot
```

Then to execute the shell program:

```
Plot 601 610
```

Here we assume that the parameters cdpmin=601, cdpmax=610 are appropriate values for the cmgs data set. Figure A.2 shows an output generated by the Plot shell program.

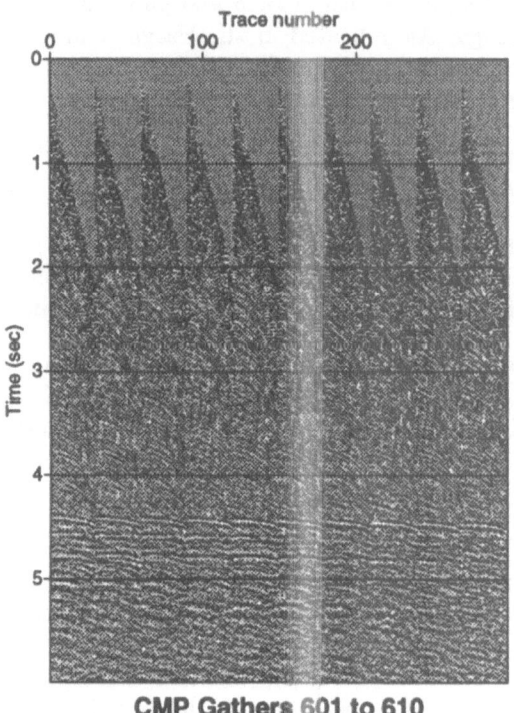

CMP Gathers 601 to 610

Fig. A.2. Output of the Plot shell program.

A typical SU processing flow. Suppose you want to use sudmofk. You've read the self-doc, but a detailed example is always welcome isn't it? The place to look is the directory su/examples. In this case, we are lucky and find the shell program, Dmo. Again, the numbers in square brackets at the end of the lines shown below are *not* part of the listing.

```
#! /bin/sh
# dmo
set -x                                                    [1]

# set parameters
input=cdp201to800                                         [2]
temp=dmocogs
output=dmocmgs
```

```
smute=1.7
vnmo=1500,1550,1700,2000,2300,2600,3000                          [3]
tnmo=0.00,0.40,1.00,2.00,3.00,4.00,6.00


# sort to common-offset, nmo, dmo, inverse-nmo, sort back to cmp
susort <$input offset cdp |                                      [4]
sunmo smute=$smute vnmo=$vnmo tnmo=$tnmo |                       [5]
sudmofk cdpmin=201 cdpmax=800 dxcdp=13.335 noffmix=4 |          [6]
sunmo invert=1 smute=$smute vnmo=$vnmo tnmo=$tnmo >$temp        [7]
susort <$temp cdp offset >$output                                [8]
```

Discussion of numbered lines:

The core of the shell program (lines 5-7) is recognized as the typical dmo process: crude nmo, dmo, and then "inverse" nmo. The dmo processing is surrounded by sorting operations (lines 4 and 8). Here is a detailed discussion of the shell program keyed to the numbers appended to the listing (see also the discussion above for the Plot shell):

1. Set a debugging mode that asks UNIX to echo the lines that are executed. You can comment this line off when its output is no longer of interest. An alternate debugging flag is set -v which echos lines as they are read by the shell interpreter. You can use both modes at once if you like.

2. This line and the next two lines set filenames that, in this case, are in the same directory as the shell program itself. Again, the reason for using parameters here is to make it easy to "clone" the shell for use with other data sets. Those of us who work with only a few data sets at any given time, find it convenient to devote a directory to a given data set and keep the shells used to process the data in that directory as documentation of the processing parameters used. (SU does not have a built-in "history" mechanism.)

3. The dmo process requires a set of velocity-time picks for the subsidiary nmo processes. Since these picks must be consistent between the nmo and the inverse nmo, it is a good idea to make them parameters to avoid editing mistakes. Again, note the format of SU parameter vectors: comma-separated strings with no spaces. The nmo program (sunmo) will give an error message and abort if the vnmo and tnmo vectors have different lengths.

4. Note that susort allows the use of *secondary* sort keys. Do not assume that a secondary field that is initially in the "right" order will remain in that order after the sort—if you care about the order of some secondary field, specify it (as this shell program does). In this line, we sort the data according to increasing offsets and then, within each offset, we sort according to increasing cdp number.

5. The forward nmo step.

6. The dmo step.

7. The inverse nmo step.

8. Sort back to cdp and have increasing offset within each cdp.

If you want to thoroughly understand this shell program, your next step is to
study the self-docs of the programs involved:

```
% sunmo

SUNMO - NMO for an arbitrary velocity function of time and CDP

sunmo <stdin >stdout [optional parameters]

Optional Parameters:
vnmo=2000          NMO velocities corresponding to times in tnmo
tnmo=0             NMO times corresponding to velocities in vnmo

   ...
```

Related shell programs are

    su/examples/Nmostack

and

    su/examples/Mig.

A.4.4 Extending SU by Shell Programming

Shell programming can be used to greatly extend the reach of SU without writing C code. See, for example, Cvstack, Filtertest, Firstbreak, and Velan in
su/examples.

It is a sad fact that the UNIX shell is not a high level programming language—
consequently, effective shell coding often involves arcane tricks. In this section,
we'll provide some useful templates for some of the common UNIX shell programming idioms.

We use Cvstack as an illustration. The core of this shell is a double loop over
velocities and cdps that produces *velocity panels*—a concept not contained in any
single SU program.

Remark: For most of us, writing a shell like Cvstack from scratch is a time-consuming affair. To cut down the development time, your authors excerpt from existing shells to make new ones even when we don't quite remember what every detail means. We suggest that you do the same!

We won't comment on the lines already explained in our previous two shell code examples (see Sections A.4.3 and A.4.3), but instead focus on the new features used in Cvstack.

```
#! /bin/sh
# Constant-velocity stack of a range of cmp gathers
# Authors: Jack, Ken
# NOTE: Comment lines preceding user input start with  #!#
set -x

#!# Set input/output file names and data parameters
input=cdp601to610
stackdata=cvstack
cdpmin=601 cdpmax=610
fold=30
space=1          # 1 null trace between panels

#!# Determine velocity sampling.
vmin=1500    vmax=3000    dv=150

### Determine ns and dt from data (for sunull)
nt='sugethw ns <$input | sed 1q | sed 's/.*ns=//''              [1]
dt='sugethw dt <$input | sed 1q | sed 's/.*dt=//''

### Convert dt to seconds from header value in microseconds
dt='bc -l <<END                                                 [2]
        scale=4
        $dt / 1000000
END'


### Do the velocity analyses.
>$stackdata  # zero output file                                 [3]
v=$vmin
while [ $v -le $vmax ]                                          [4]
do
        cdp=$cdpmin
        while [ $cdp -le $cdpmax ]                              [5]
        do
                suwind <$input \                               [6]
                        key=cdp min=$cdp max=$cdp count=$fold |
                sunmo cdp=$cdp vnmo=$v tnmo=0.0 |
                sustack >>$stackdata
                cdp='bc -l <<END                               [7]
                        $cdp + 1
END'
        done
```

```
            sunull ntr=$space nt=$nt dt=$dt >>$stackdata          [8]
v='bc -1 <<END
            $v + $dv
END'
done


### Plot the common velocity stacked data
ncdp='bc -1 <<END
            $cdpmax-$cdpmin+1
END'
f2=$vmin
d2='bc -1 <<END
            $dv/($ncdp + $space)                                  [9]
END'

sugain <$stackdata tpow=2.0 |

suximage perc=99 f2=$f2 d2=$d2 \
            title="File: $input  Constant-Velocity Stack " \
            label1="Time (s)"  label2="Velocity (m/s)" &

exit                                                              [10]
```

Discussion of numbered lines:

1. This elaborate construction gets some information from the first trace header
   of the data set. The program sugethw lists the values of the specified keys in
   the successive traces. For example,

```
    % suplane | sugethw tracl ns
      tracl=1              ns=64

      tracl=2              ns=64

      tracl=3              ns=64

      tracl=4              ns=64

      tracl=5              ns=64

      tracl=6              ns=64

      . . .
```

Although sugethw is eager to give the values for every trace in the data set,
we only need it once. The solution is to use the UNIX stream editor (sed). In
fact, we use it twice. By default, sed passes along its input to its output. Our
first use is merely to tell sed to quit after it puts the first line in the pipe. The
second pass through sed strips off the unwanted material before the integer.

In detail, the second **sed** command reads: replace (or substitute) everything up to the characters **ns=** with nothing, i.e., delete those characters.

2. We are proud of this trick. The Bourne shell does not provide floating point arithmetic. Where this is needed, we use the UNIX built-in **bc** calculator program with the "here document" facility. Here, we make the commonly needed conversion of sampling interval which is given in micro-seconds in the SEG-Y header, but as seconds in SU codes. Note carefully the *back*quotes around the entire calculation—we assign the result of this calculation to the shell variable on the left of the equal sign, here dt. The calculation may take several lines. We first set the number of decimal places with **scale=4** and then do the conversion to seconds. The characters END that follow the here document redirection symbol **<<** are arbitrary, the shell takes its input from the text in the shell file until it comes to a line that contains the same characters again. For more information about bc:

```
% man bc
```

3. As the comment indicates, this is a special use of the output redirection symbol that has the effect of destroying any pre-existing file of the same name or opening a new file with that name. In fact, this is what **>** always does as its first action—it's a dangerous operator! If you intend to *append*, then, as mentioned earlier, use **>>**.

4. This is the outer loop over velocities. Another warning about spaces—the spaces around the bracket symbols are essential.

   Caveat: The bracket notation is a nice alternative to the older clunky **test** notation:

```
while test $v -le $vmax
```

   Since the bracket notation is not documented on the typical **sh** manual page, we have some qualms about using it. But, as far as we know, all modern **sh** commands support it—please let us know if you find one that doesn't.

   WARNING! OK, now you know that there is a UNIX command called **test**. So don't use the name "test" for one of your shell (or C) programs—depending on your $PATH setting, you could be faced with seemingly inexplicable output.

5. This is the inner loop over cdps.

6. Reminder: No spaces or tabs after the line continuation symbol!

7. Notice that we broke the nice indentation structure by putting the final END against the left margin. That's because the **sh** manual page says that the termination should contain only the END (or whatever you use). In fact, most versions support indentation. We didn't think the added beautification was

worth the risk in a shell meant for export. Also note that we used bc for an integer arithmetic calculation even though integer arithmetic is built into the Bourne shell—why learn two arcane rituals, when one will do? See man expr, if you are curious.

8. sunull is a program I (Jack) wrote to create all-zero traces to enhance displays of the sort produced by Cvstack. Actually, I had written this program many times, but this was the first time I did it on purpose. (Yes, that was an attempt at humor.)

9. An arcane calculation to get velocity labeling on the trace axis. Very impressive! I wonder what it means? (See last item.)

10. The exit statement is useful because you might want to save some "spare parts" for future use. If so, just put them after the exit statement and they won't be executed.

Figure A.3 shows an output generated by Cvstack.

A.4.5 Some Core Programs

Reading the self-documentation and trying out the following SU programs will give you a good start in learning SU.

Examining the Trace Headers.

surange — print minimum and maximum values of trace header fields

sugethw — print values of selected header fields

suascii — print header and data values

suxedit — interactively examine headers and traces

Some Common Processing Programs.

suacor — compute autocorrelations

sufilter — multipurpose zero phase filter (includes bandpass)

sugain — gain (with lots of options)

sumute — zero samples before a time that depends on offset

sunmo — normal-moveout correction

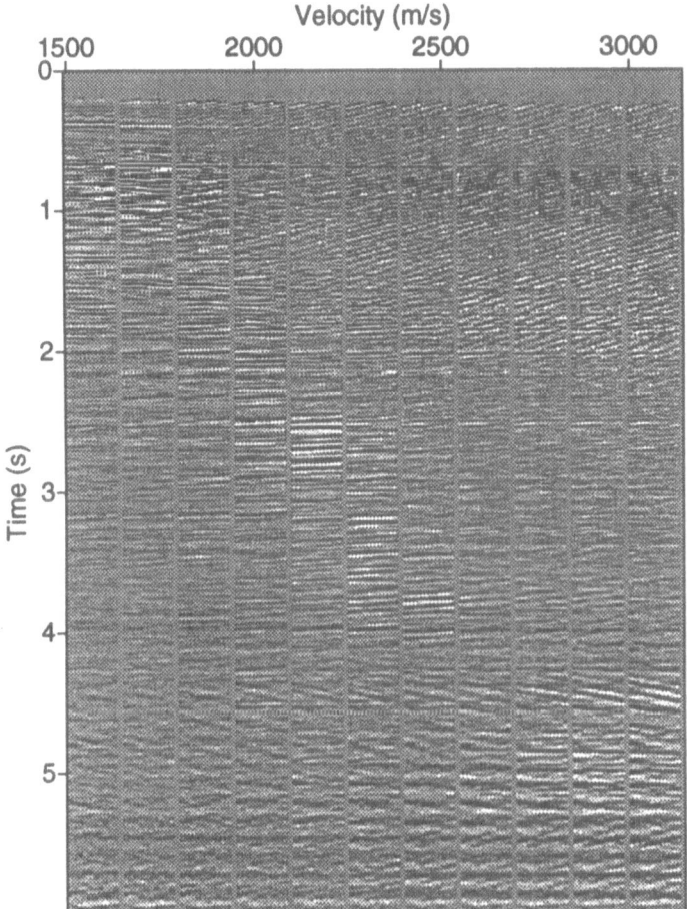

File: cdp601to610 Constant-Velocity Stack

Fig. A.3. Output of the Cvstack shell program.

supef — prediction error filtering

susort — sort traces by values of trace header fields

sustack — stack (sum) traces

suvelan — velocity analysis

suwind — window (i.e., get a subset of) traces

Some Common Plotting Programs.

suximage — gray scale X Windows plotting

suxwigb — bit mapped wiggle trace X Windows plotting

supsimage — gray scale PostScript plotting

supswigb — bit mapped wiggle trace PostScript plotting

A.4.6 A Brief Tour of the Source Directories

The SU software is a layered product. The layers correspond to the following directories:

cwp Library of scientific routines (e.g. fft routines) written in "vanilla" C. Utility mains and shells.

par Library supporting the CWP programming style (i.e., self-doc, error reporting, parameter passing). Mains that use (only) these facilities. Shells for maintaining the online documentation database.

su Seismic processing codes that use the SEG-Y trace structure. Subroutines that manage this structure. Codes that buffer the generic graphics routines listed below. Shells that provide backward compatibility with earlier releases.

graphics libraries

1. **psplot**—PostScript graphics:

   a) pscontour: contour plots

   b) pscube: 3D data cube

   c) psgraph: curve plotting

   d) psimage: raster plotting

    e) psmovie: supports frames

    f) pswigb: bit mapped wiggle traces (fast)

    g) pswigp: polygon wiggle traces (slow)

    h) PostScript support programs

2. xplot—xlib based X Windows graphics

    a) ximage: raster plotting

    b) xwigb: bit mapped wiggle traces

    c) X Windows support programs

3. **Xtcwp—toolkit based X Windows graphics**

    a) **xgraph: curve plotting**

    b) **xmovie: supports frames**

    c) **X Windows resource files**

These are only the highlights. If you intend to add your own C mains to the package, it is worthwhile spending a few hours browsing through the source code.

A.5 Frequently Asked Questions

This section addresses questions often asked by new SU users. Some answers refer to the directory CWPROOT. We use this symbolic name for the directory that contains the CWP/SU source code, include files, libraries, and executables. You are asked to specify this directory name during the SU installation procedure.

A.5.1 Installation Questions

Complete information about the installation process is found in the README files supplied with the distribution. Here we discuss only some commonly found installation problems.

Question A.5.1. *I get error messages about missing* **fgetpos** *and* **fsetpos** *routines, even though I am using the* GCC *compiler. How do I get around this problem?*

Answer A.5.1. We've seen this problem most often with older SUN OS 4.xx (pre-SOLARIS). These SUN systems may not have the `fgetpos` and `fsetpos` subroutines defined. Since these two routines are not currently used in the SU package, the easiest fix when these two functions are missing functions is to comment out references to the functions `efgetpos` and `efsetpos` in CWPROOT/src/par/include/par.h and CWPROOT/src/par/lib/subcalls.c. See also the answer to Question A.5.3.

Question A.5.2. *I get error messages about missing* `strtoul`, *and/or* `strerror` *routines, even though I am using the* GCC *compiler. How do I get around this problem?*

Answer A.5.2. Again, this is most often seen with the older SUN OS. To fix the problem of a missing `strtoul` routine, replace:
CWPROOT/src/par/lib/atopkge.c
with the file:
CWPROOT/src/par/lib/Portability/atopkge.c.

For a missing `strerror` routine, replace:
CWPROOT/src/par/lib/errpkge.c
with the file:
CWPROOT/src/par/lib/Portability/errpkge.c.

Each of these replacements, of the regular file by the `Portability` version, entails a small loss of functionality. See also the answer to Question A.5.3.

Question A.5.3. *Why do I get missing subroutine messages about* ANSI C *routines? Isn't the* GCC *compiler supposed to be an* ANSI *compiler?*

Answer A.5.3. The GCC compiler is just that, a compiler. It draws on the libraries that are present on the machine. If the GNU libraries (this is the "glibc" package) have not been installed, then the GCC compiler will use the libraries that are native to the machine you are running on. Because the four routines listed above are not available in the SUN 4. OS, GCC does not recognize them. However, installing the GNU libraries will make the GCC compiler behave as a full ANSI C compiler.

A.5.2 Data Format Questions

In this section, we address questions about converting data that are in various formats into SU format.

Question A.5.4. *What is the data format that* SU *programs expect?*

Answer A.5.4. The SU data format is based on the SEG-Y format. The SU format consists of data traces each of which has a header. The SU trace header is identical to SEG-Y trace header. Both the header and the trace data are written in the native binary format of your machine.

Caution: The optional fields in the SEG-Y trace header are used for different purposes at different sites. SU itself makes use of certain of these fields. Thus, you may need to use segyclean—see the answer to Question A.5.6. SU format does not have the binary and ebcdic tape headers that are part of the SEG-Y format.

After installing the package, you can get more information on the SEG-Y/SU header by typing:

```
% sukeyword -o
```

This lists the include file segy.h that defines the SU trace header.

Question A.5.5. *Is there any easy way of adding necessary SEG-Y information to our own modeled data to prepare our data for processing using the SU package?*

Answer A.5.5. It depends on the details of how your data was written to the file:

1. If you have a 'datafile' that is in the form of binary floating point numbers of the type that would be created by a C program, then use suaddhead to put SU (SEG-Y) trace headers on the data. Example:

   ```
   % suaddhead < datafile  ns=N_SAMP > data.su
   ```

   Here, N_SAMP is the (integer) number of samples per trace in the data.

2. If your data are Fortran-style floats, then you would use:

   ```
   % suaddhead < datafile ftn=1 ns=NS > data.su
   ```

   See also, Question A.5.9.

3. If your data are ASCII, then use:

   ```
   % a2b n1=N1 < data.ascii | suaddhead ns=NS > data.su
   ```

   Here N1 is the number of floats per line in the file data.ascii.

4. If you have some other data type, then you may use:

   ```
   % recast < data.other in=IN out=float |
             suaddhead ns=NS > data.su
   ```

where IN is the type (int, double, char, etc...)

For further information, consult the self-docs of the programs suaddhead, a2b, and recast.

Question A.5.6. *I used* segyread *to read a SEG-Y tape. Everything seems to work fine, but when I plot my data with* suximage, *the window is black. What did I do wrong?*

Answer A.5.6. When you read an SEG-Y tape, you need to pipe the data through segyclean to zero the optional SEG-Y trace header field. If the SU programs see nonzero values in certain parts of the optional field, they try to display the data as "nonseismic data," using those values to set the plot parameters.

Question A.5.7. *I am trying to plot data with the* pswigb *(or* pswigp, *or* xwigb, *or ...) program. I know that I have data with* n1=NSAMP *and* n2=NTRACES, *but when I plot, I find that I have to set* n1=NSAMP+60 *for the plot to look even remotely correct. Why is this?*

Answer A.5.7. It is likely that you are trying to plot with the wrong tool. The input data format of the programs, pswigb, pswigp, pscontour, pscube, psmovie, xwigb, xgraph, and xmovie, expect data to consist of simple floating point numbers. If your data are SU data (SEG-Y) traces, then there is an additional header at the beginning of each trace, which, on most computer architectures, is the same number (240) of bytes as the storage for 60 floats.

To plot these data, use respectively: supswigb, supswigp, supscontour, supscube, supsmovie, suxwigb, suxgraph, or suxmovie.

Also, it is not necessary to specify the dimensions of the data for these latter programs. The su-versions of the codes determine the necessary information from the appropriate header values. (In fact, that is *all* they do—the actual graphics is handled by the version without the su prefix.)

Question A.5.8. *I want to check the size of a file to see if it has the right number of values, but I am not sure how to take the header into account. How is this done?*

Answer A.5.8. If the file consists of simple floating point numbers, then the size in bytes equals the size of a float times the number of samples (SIZE=4*N_SAMP). The SU data (SEG-Y traces) also have a header (240 bytes per trace) giving the total number of bytes as:
(240+4*N_SAMP)*N_TRACES.
The byte count computed in this way is the number that the UNIX command ls -l shows.

Caveats: The above calculations assume that you have the conventional architecture and that the header definition in `segy.h` has not been altered. Watch out as machines with 64 bit word size become common!

Question A.5.9. *I have some data in Fortran form and tried to convert it to* SU *data via the following:*

```
% suaddhead < data.fortran ns=N_SAMP ftn=1 > data.su
```

but this did not work properly. I am sure that my fortran data are in unformatted binary floats. What should I do?

Answer A.5.9. There are different ways of interpreting the term "unformatted" with regard to fortran data. Try:

```
% ftnstrip < data.fortran | suaddhead ns=N_SAMP > data.su
```

The program `ftnstrip` can often succeed in converting your fortran data into C-like binary data, even when the `ftn=1` option in `suaddhead` fails.

Question A.5.10. *I just successfully installed the* CWP/SU *package, but when I try to run the demo scripts, I get many error messages describing programs that the shell script cannot find. How do I fix this?*

Answer A.5.10. You need to put CWPROOT/bin in your shell PATH, where CWPROOT is /your/root/path that contains the CWP/SU source code, include files, libraries, and executables. This is done in your `.cshrc` file if you run under `csh` or `tcsh`. In Bourne shell (sh), Born Again shell (bash), or Korn shell (ksh) the PATH variable is in your `.profile` file. You also need to type

```
% rehash
```

under `.csh` or `.tcsh`, if you have not relogged since you compiled the codes.

A.5.3 Tape Reading and Writing

This section contains frequently asked questions about reading and writing SEG-Y tapes with SU.

Tape reading/writing is more of an art than a science. Here are a few tips.

1. Make sure your tape drive is set to be variable block length. If you are on an IBM RS6000, this means you will need to use smit to set blocksize=0 on your tape device. Having the tape drive set to some default constant blocksize (say blocksize=1024 or 512) will foil all attempts to read an SEG-Y tape.

2. To read multiple tape files on a tape, use the non rewinding device. On an RS6000 this would be something like /dev/rmtx.1, see man mt for details.

3. If this still doesn't work, then try:

```
% dd if=/dev/rmtx of=temps bs=32767 conv=noerror
```

Here, /dev/rmtx (not the real name of the device, it varies from system to system) is your regular (rewinding) tape device. In the option, bs=32767, we gave the right blocksize ($2^{16} + 1$) for an IBM/RS6000. Try bs=32765 ($2^{16} - 1$) on a SUN. This will dump the entire contents of the tape onto a single file.

Question A.5.11. *How do I write multiple SEG-Y files onto a tape?*

Answer A.5.11. Here is a shell script for writing multiple files on a tape:

```
#! /bin/sh

DEV=/dev/nrxt0   # non rewinding tape device

mt -f $DEV rewind

j=0
jmax=40

while test "$j" -ne "$jmax"
do
     j='expr $j + 1'
     echo "writing tape file  $j"
     segywrite tape=$DEV bfile=b.$j hfile=h.$j verbose=1 buff=0 |
< ozdata.$j
done

exit 0
```

A.5.4 General

This section addresses general questions about the SU package.

Question A.5.12. *Why are CWP/SU releases given by integers (22, 23, 24, etc...) instead of the more familiar decimal release numbers (1.1, 1.3, etc...)?*

Answer A.5.12. The CWP/SU release numbers are chosen to correspond to the SU NEWS email messages. The individual codes in the package have traditional decimal release numbers (assigned by RCS), but these are all different. The package changes in incremental, but non-uniform ways, so the standard notation seems inappropriate. However, the user may view 24 to be 2.4. We may adopt this convention in the future.

Remark: In the early days, we *did* use RCS to simultaneously update all the codes to 2.1, 3.1, ... This practice died a natural death somewhere along the way.

Question A.5.13. *How often are the codes updated?*

Answer A.5.13. The CWP/SU package is updated at roughly 3 month intervals. We mail announcements of these releases to all known users. Since we do not provide support for outdated versions, we urge you to remain current.

Question A.5.14. *I have a complicated collection of input parameters for a CWP/SU program. I want to run the command from the command line of a terminal window, but I don't want to retype the entire string of input parameters. what do I do?*

Answer A.5.14. CWP/SU programs that take their input parameters from the command line also have the feature of being able to read from a "parameter file." This is invoked by setting the parameter par=parfile, where parfile is a file containing the desired commandline string.

For example:

```
suplane ntr=20 nt=40 dt=.001 | ...
```

is completely equivalent to the command:

```
suplane par=parfile | ...
```

The string

```
ntr=20 nt=40 dt=.001
```

is contained in 'parfile.'

A.6 How to Write an SU Program

A.6.1 A Template SU Program

Although variations are usually needed, a template for a typical SU program looks like the program listing below (we excerpted lines from the program sumute to build this template). The numbers in square brackets at the end of the lines in the listing are not part of the listing—we added them to facilitate discussion of the template. The secret to efficient SU coding is finding an existing program similar to the one you want to write. If you have trouble locating the right code or codes to "clone," ask us—this can be the toughest part of the job!

```
/* SUMUTE: $Revision: 1.3 $ ; $Date: 94/10/13 09:17:39 $    */ [1]

#include "su.h"                                                  [2]
#include "segy.h"

/*********************** self documentation ****************/ [3]
char *sdoc[] = {
"                                                            ",
" SUMUTE - ......                                            ",
"                                                            ",
" sumute <stdin >stdout                                      ",
"                                                            ",
" Required parameters:                                       ",
"        none                                                ",
"                                                            ",
" Optional parameters:                                       ",
"        ...                                                 ",
"                                                            ",
" Trace header fields accessed: ns                           ",
" Trace header fields modified: none                         ",
"                                                            ",
NULL};
/*************** end self doc ****************************/

/* Credits:
 *
 *        CWP: Jack Cohen, John Stockwell
 */


segy tr;                                                        [4]

main(int argc, char **argv)
{
        int ns;                 /* number of samples    */      [5]
        ...
```

```
/* Initialize */
initargs(argc, argv);                                    [6]
requestdoc(1);                                           [7]


/* Get parameters */
if (!getparint("ntaper", &ntaper))        ntaper = 0;    [8]


/* Get info from first trace */
if (!gettr(&tr)) err("can't read first trace");          [9]
if (!tr.dt) err("dt header field must be set");         [10]

/* Loop over traces */
do {                                                    [11]
        int nt     = (int) tr.ns;                       [12]

        if (below == 0) {                               [13]
                nmute = NINT((t - tmin)/dt);
                memset((void *) tr.data, (int) '\0',
                        nmute*FSIZE);
                for (i = 0; i < ntaper; ++i)
                        tr.data[i+nmute] *= taper[i];
        } else {
                nmute = NINT((nt*dt - t)/dt);
                memset((void *) (tr.data+nt-nmute),
                                (int) '\0', nmute*FSIZE);
                for (i = 0; i < ntaper; ++i)
                        tr.data[nt-nmute-1-i] *= taper[i];
        }
        puttr(&tr);                                     [14]
} while (gettr(&tr));                                    [15]

        return EXIT_SUCCESS;                            [16]
}
```

Discussion of numbered lines:

1. We maintain the internal versions of the codes with the UNIX utility RCS. This item shows the string template for RCS.

2. The file su.h includes (directly or indirectly) all our locally defined macros and prototypes. The file segy.h has the definitions for the trace header fields.

3. The starred lines delimit the "self-doc" information—include them exactly as you find them in the codes since they are used by the automatic documentation shells. The style of the self-doc shown is typical except that often additional usage information is shown at the bottom and, of course, often there are more options. Look at some existing codes for ideas.

4. This is an external declaration of an SU (SEG-Y) trace buffer. It is external to avoid wasting stack space.

5. We usually describe the global variables at the time of declaration. Examine codes related to yours to increase consistency of nomenclature (there is no official SU naming standard).

6. The **initargs** subroutine sets SU's command line passing facility (see page 227).

7. The **requestdoc** subroutine call specifies the circumstances under which self-doc will be echoed to the user. The argument '1' applies to the typical program that uses only standard input (i.e. <) to read an SU trace file. Use '0' for codes that create synthetic data (like **suplane**) and '2' for codes that require two input files (we could say "et cetera," but there are no existing SU mains that require *three* or more input files).

8. This is typical code for reading 'parameters from the command line. Interpret it like this: "If the user did not specify a value, then use the default value." The subroutine must be type-specific, here we are getting an *integer* parameter.

9. Read the first trace, exit if empty. The subroutine **fgettr** "knows about" the SU trace format. Usually the trace file is read from standard input and then we use **gettr** which is a macro based on **fgettr** defined in **su.h**. Note that this code implies that the first trace is read into the trace buffer (here called **tr**), therefore we will have to process this trace before the next call to **fgettr**.

10. We've read that first trace because, we need to get some trace parameters from the first trace header. Usually these are items like the number of samples (**tr.ns**) and/or the sampling interval (**tr.dt**) that, by the SEGY-Y standard, are the same for all traces.

11. Since the first trace has been (typically) read before the main processing loop starts, we use a "do-while" that reads a new trace at the *bottom* of the loop.

12. We favor using *local* variables where permitted.

13. This is the seismic algorithm–here incomplete. We've left in some of the actual **sumute** code because it happens to contains lines that will be useful in the new code, we'll be writing below. You may want to call a subroutine here to do the real work.

14. **fputtr** and **puttr** are the output analogs of **fgettr** and **gettr**.

15. The loop end. **gettr** returns a 0 when the trace file is exhausted and the processing then stops.

16. This is an ANSI-C macro conventionally used to indicate successful program termination.

A.6.2 Writing a New Program: suvlength

A user asked about SU processing for variable length traces. At his institute, data are collected from time of excitation to a variable termination time. The difficulty is that SU processing is based on the SEG-Y standard which mandates that all traces in the data set be of the same length. Rather than contemplating changing all of SU, it seems to us that the solution is to provide a program that converts the variable length data to fixed length data by padding with zeroes where necessary at the end of the traces—let's name this new program suvlength. We can make the length of the output traces a user parameter. If there is a reasonable choice, it makes sense to provide a default value for parameters. Here, using the length of the first trace seems the best choice since that value can be ascertained before the main processing loop starts.

So far, so good. But now our plan runs into a serious snag: the fundamental trace getting facility, gettr, itself assumes fixed length traces (or perhaps we should say that gettr deliberately enforces the fixed length trace standard). But, if you think about it, you'll realize that gettr itself has to take special measures with the *first* trace to figure out its length. All we have to do is make a new trace getting routine that employs that first trace logic for *every* trace! Fortunately, we don't even have to do that since the same problem arose a few years ago and we wrote fvgettr at that time for the requirements of the suvcat program. So as a first draft solution, we'll just copy in fvgettr as an in-code subroutine for our new program. Let's begin converting our above template into the new suvlength code:

```
/* SUVLENGTH: $Revision: 1.3 $ ; $Date: 94/10/13 09:17:39 $ */

#include "su.h"
#include "segy.h"

/*********************** self documentation **********************/
char *sdoc[] = {
"                                                                ",
" SUVLENGTH - Adjust variable length traces to common length     ",
"                                                                ",
" suvlength <variable_length_traces >fixed_length_traces          ",
"                                                                ",
" Required parameters:                                            ",
"         none                                                    ",
"                                                                ",
" Optional parameters:                                            ",
"         ns      output number of samples (default: 1st trace ns)",
"                                                                ",
" Trace header fields accessed: ns                                ",
" Trace header fields modified: ns                                ",
"                                                                ",
NULL};
```

```
/**************** end self doc ********************************/

/* Credits:
 *
 *          CWP: Jack Cohen, John Stockwell
 */

/* prototype */
int fvgettr(FILE *fp, segy *tp);

segy tr;

main(int argc, char **argv)
{
        int ns;          /* number of samples on output traces  */

        /* Initialize */
        initargs(argc, argv);
        requestdoc(1);

        /* Get parameters */
        ...

        /* Get info from first trace */
        ...

        ...

        return EXIT_SUCCESS;                                    [16]
}

/* fvgettr code from suvcat goes here */
        ...
```

Now we run into a small difficulty. Our only parameter has a default value that is obtained only after we read in the first trace. The obvious solution is to reverse the parameter getting and the trace getting in the template. Thus we resume:

```
        /* Get info from first trace and set ns */
        if (!fvgettr(stdin, &tr))  err("can't get first trace");
        if (!getparint("ns", &ns))     ns = tr.ns;

        /* Loop over the traces */
        do {
                int nt = tr.ns;
```

Now comes the actual seismic algorithm—which is rather trivial in the present case: add zeroes to the end of the input trace if the output length is specified greater than the input length. We could write a simple loop to do the job, but the task is done most succinctly by using the ANSI-C routine memset. However, we confess that unless we've used it recently, we usually forget how to use this routine. One solution is to cd to the su/main directory and use grep to find other uses of memset. When we did this, we found that sumute had usage closest to what we needed and that is why we started from a copy of that code. Here is the complete main for suvlength:

```
/* SUVLENGTH: $Revision: 1.3 $ ; $Date: 94/10/13 09:17:39 $    */

#include "su.h"
#include "segy.h"

/*********************** self documentation *******************/
char *sdoc[] = {
"                                                            ",
" SUVLENGTH - Adjust variable length traces to common length ",
"                                                            ",
" suvlength <vdata >stdout                                   ",
"                                                            ",
" Required parameters:                                       ",
"         none                                               ",
"                                                            ",
" Optional parameters:                                       ",
"         ns      output number of samples (default: 1st trace ns)",
"                                                            ",
" Trace header fields accessed:  ns                          ",
" Trace header fields modified:  ns                          ",
"                                                            ",
NULL};
/*************** end self doc *******************************/

/* Credits:
 *       CWP: Jack Cohen, John Stockwell
 */

/* prototype */
int fvgettr(FILE *fp, segy *tp);

segy tr;

main(int argc, char **argv)
{
        int ns;                 /* samples on output traces      */

        /* Initialize */
        initargs(argc, argv);
        requestdoc(1);

        /* Get info from first trace */
        if (!fvgettr(stdin, &tr))  err("can't get first trace");
        if (!getparint("ns", &ns))    ns = tr.ns;
```

```
        /* Loop over the traces */
        do {
                int nt = tr.ns;

                if (nt < ns) /* pad with zeros */
                        memset((void *)(tr.data + nt), '\0',
                        (ns-nt)*FSIZE);
                tr.ns = ns;
                puttr(&tr);
        } while (fvgettr(stdin, &tr));

        return EXIT_SUCCESS;
}


#include "header.h"

/* fvgettr - get a segy trace from a file by file pointer (nt
 *              can vary)
 *
 * Returns:
 *        int: number of bytes read on current trace (0 after
 *              last trace)
 *
 * Synopsis:
 *        int fvgettr(FILE *fp, segy *tp)
 *
 *
 * Credits:
 *        Cloned from .../su/lib/fgettr.c
 *
 */


int fvgettr(FILE *fp, segy *tp)
    ...
```

Of course, now that fvgettr has been used in two codes, it should be extracted as a library function and we should make a convenience macro vgettr for the case of standard input. But these are secondary considerations that don't arise for most applications.

For any new SU code, one should provide an example shell program to show how the new code is to be used. Here is such a program for X Windows graphics:

```
#! /bin/sh
# Trivial test of suvlength with X Windows graphics
# Use same graphics set-up as in demos (e.g.
# demos/Sorting_Traces/Demo/Xsort)

GRAPHER=xgraph
```

```
IMAGER=suximage
WIGGER=suxwigb
WIDTH=700
HEIGHT=900
WIDTHOFF=50
HEIGHTOFF=20

>tempdata
>vdata
suplane >tempdata #default is 32 traces with 64 samples per trace
suplane nt=72 >>tempdata
suvlength <tempdata ns=84 |
sushw key=tracl a=1 b=1 >vdata

# Plot the data
$WIGGER <vdata \
        perc=99 title="suvlength test"\
        label1="Time (sec)" label2="Traces" \
        -geometry ${WIDTH}x${HEIGHT}+${WIDTHOFF}+${HEIGHTOFF} \
        wbox=$WIDTH hbox=$HEIGHT xbox=$WIDTHOFF ybox=$HEIGHTOFF &

# Remove #comment sign on next line to test the header
#sugethw <vdata tracl ns | more

# Clean up
rm tempdata vdata
```

and here is the PostScript equivalent:

```
#! /bin/sh
# Trivial test of suvlength with PostScript graphics
# Use same graphics set-up as in demos (e.g.
# demos/Sorting_Traces/Demo/PSsort)
# set PostScript Previewer here if environment
# variable PSPREVIEWER not set
VIEWER=$PSPREVIEWER

GRAPHER=psgraph
IMAGER=supsimage
WIGGER=supswigp

>tempdata
>vdata
suplane >tempdata  # default is 32 traces with 64
                   # samples per trace
suplane nt=72 >>tempdata
suvlength <tempdata ns=84 |
sushw key=tracl a=1 b=1 >vdata

# Plot the data
```

```
$WIGGER <vdata perc=99 title="suvlength test"\
        label1="Time (sec)" label2="Traces" | $VIEWER

# Remove #comment sign on next line to test the header
#sugethw <vdata tracl ns | more

# Clean up
rm tempdata vdata
exit
```

B. SUB User's Guide

Martin L. Smith
New England Research, Inc.
76 Olcott Drive
White River Junction, Vermont 05001
USA
martin@ner.com

SUB combines a complete, simple procedural language with convenient support for reading, manipulating, and writing SEG-Y binary trace record data streams. SUB makes it easy for the user to inspect, report, and modify the contents of a SEG-Y data stream by writing simple programs in SUB's internal language. SUB also supports efficient trace-oriented manipulation which makes it possible to perform sophisticated time series processing. SUB is freely available via ftp from the Center for Wave Phenomena, as explained in the previous appendix.

This appendix describes describes SUB (v. 0.9). SUB (v. 0.9) is the first public release and is a *beta* release. Although much of the code has seen substantial testing, there are many new features which probably contain many new bugs. This is also the first version of the **User's Guide**. Address questions, comments, and complaints to Martin Smith at the address above.

B.1 Some SUB Examples

The quickest way to get an idea of what SUB is useful for is to examine some simple uses. If we execute

```
sub stdscript1 < sample.segy
```

where `sample.segy` is a sample SEG-Y data set included with the SUB release and `stdscript1` is a file containing the text

```
func Begin()
{       Records = 0;
        sumavgsq = 0;
}
```

```
func OnTrace()
{       ++Records;
        sumavgsq = sumavgsq +
sum(Tr.trace * Tr.trace)/length(Tr.trace);
}
```

```
func End()
{       auto avgsq;
        if(Records == 0) exit(0);
        avgsq = sumavgsq/Records;
        print("Records: ", Records, "\\n");
        print("Average squared amplitude: ", avgsq, "\\n");
}
```

then we get the output (to standard error)

```
Records: 40
Average squared amplitude: 1.66367e-05
```

This example shows several important characteristics of a SUB script.

– Scripts structured by functions

The contents of stdscript1 specify a program which SUB interprets and executes. We call such a program a *script*. This script consists of several function definitions: Begin, OnTrace, and End. These are automatically invoked (if they exist) by the interpreter at different points in the program's life:

1. Begin() is invoked before anything has been read from standard input.

2. OnTrace() is invoked each time a SEG-Y trace record is read. Further, the contents of the record are made available in the global variable Tr.

3. End() is invoked after an end-of-file has been encountered on standard input.

– Convenient data element access

The script shows how components of the SEG-Y trace record are invoked as *members* of the trace record. In this example Tr.trace is the trace member of the SEG-Y trace record contained in Tr. Accessing the member Tr.dt would return the (integer) sample interval in microseconds. (See the end of this appendix for a complete list of SEG-Y trace record members.)

– Convenient trace processing operations

When we computed the sum of the squared trace values, we exploited SUB's support for vector-based operations. In this particular instance where Tr.trace is a vector of n elements, the expression

```
Tr.trace * Tr.trace
```

results in a new vector of n elements, each element of which is the square of the corresponding element in Tr.trace. (This definition is not the dot product of two vectors; it is the element-by-element product. To compute the dot product of two vectors, a and b, use sum(a * b).)

In the above SUB scanned the incoming SEG-Y stream and extracted information from the traces contained in it. We could just as easily have modified the values in the stream and passed the new values onward in the data steam. Here is a script that normalizes each trace so that its average squared sample value is one and then passes the modified data onward.

```
func OnTrace()
{       rmsf = 1.0 / sqrt( sum(Tr.trace * Tr.trace)
                    /length(Tr.trace));
        Tr.trace = rmsf * Tr.trace;
        fputrb(stdout, Tr);
}
```

If we execute

```
sub leadscript1 < sample.segy | sub stdscript1
```

then we get the output (to standard error)

```
Records: 40
Average squared amplitude: 1
```

In this example the (modified) SEG-Y data form the first script (**leadscript1**) is piped on a second copy of SUB running the script form the first example. The text output we see at the terminal comes from the second script.

B.2 SUB Syntax

This section summarizes the properties of SUB's *C-like* scripting language. SUB gives special attention to arrays, and particularly arrays of floating-point values. That topic is so important that we've moved it to a section of it's own.

B.2.1 Names

Variable and function names are limited by the same restrictions as C. Names are case-independent. The only time SUB retains alphabetic case is in the contents of *strings*.

B.2.2 Variables and Types

SUB variables are always regular (named) variables. Variables are global by default. They are created simply by being referenced.

The type of a variable is simply the type of whatever value was last assigned to it. A variable may assume many different types during a computation. The possible contents of a variable are:

− Nothing

  A value and a type; it is the value of a variable which has not been given a value

− double

  A double-precision floating-point value

− string

  A sequence of ASCII characters (internally terminated by an ASCII zero)

− RBlock

  A SEG-Y binary trace record, consisting of a trace header and a trace time series

− stream

  A pointer to an input or output stream (such as stdout or a value returned by fopen() or popen())

— array

An array of values of any of these types, or

— fVector

A special type of array containing only floating-point values. The values in an fVector are stored as single-precision values.

B.2.3 Functions

Function definitions begin with the keyword func. Functions must be defined before they appear in the script. Functions may have arguments and local variables. Here's a simple function definition:

```
//
//   return the dot product of two one-dimensional arrays.
//   Using explicit indexing, as we do here, is very slow
//   compared to vector manipulation.
//
  func dot(a, b) {
          auto sum, i;
          sum = 0;
          for(i = 0; i < min(length(a), length(b)); i++)
                  sum = sum + a[i] * b[i];
          return sumsq;
  }
```

Functions in SUB are re-entrant (they may call themselves either directly or indirectly). If the return statement does not have an argument, or if there is no return statement, the function returns the value Nothing.

B.2.4 Comments

SUB supports three styles of comments:

— shell: from # to end-of-line,

— C: everything from * to */, and

— C++: everything from // to end-of-line.

Comments cannot be nested. Don't look for trouble.

Note: Please avoid using shell-style comments in the future; that style will be declared illegal as soon as I can get away with it so that we can pass scripts through a preprocessor.

B.2.5 Statements

SUB uses semicolons and curly braces in the same way C does. All simple statements should end with semicolons. Block statements should be wrapped in curly braces (and do not have a teminating semicolon).

B.2.6 Operators

Numeric Operators. The operators '=', '*', '/', '-', and '+' are available for operating on numeric values. Also available are '%' (for modulo) and '^' (for raised-to-the-power-of). The increment and decrement operators, '++' and '--', are available in both their postfix and prefix forms and work just as in C.

String Operators. Only '=' and '+' are available for strings, with '+' denoting concatenation. Strings cannot be mixed with other types at present; thus the expression "3" + 7 is illegal (and will be caught by the compiler whenever possible and at run-time otherwise).

Comparison Operators. The comparison operators '==', '>', '<', '>=', '<=', and '!=' are only implemented for numeric comparisons. No other types may be used as arguments to these operators at present. (I'll try to fix this soon so we at least have some string comparison support.)

Logical Operators. We currently support '&' and '&&' for *and*, '|' and '||' for *or*, and '!' for *not*. At present all of the operands in a logical expression are evaluated (which is not generally true in C). For now, don't do anything that depends on partial evaluation. (If you don't understand what this is about, you're safe from this error.)

There is no comma operator. There are no pointers and there are no address-related operators.

B.2.7 Control Flow

SUB supports

— for(*test*)

— if(*test*) ...else

— while(*test*)

— for(*init; test; incr*)

These can be combined and nested just as in C. We **do not** yet support:

— break

— continue

— switch(*value*){ ...}

There is no **goto** statement and there never will be.

B.2.8 Storage Management

Storage is managed automatically and all assignments are *assignments by value*, which means that a new copy of the data is used. There is no notion of pointers or of explicit memory management by the user.

B.3 SUB Arrays

B.3.1 Arrays

An array is a set of elements accessed *via* subscripts. Array subscripts start at zero (like C) and go up (although you are free to pretend that they start at one). Subscripts use [] and multiple subscripts require multiple [] pairs: x[i][j][7].

SUB supports the special index value $, which equals the index of the last element in the array. Expressions like

```
a[$+1] = 77;
```

are legal and add an element to the end of an array. An array's elements need not all have the same type. An array element may itself contain an array.

Referring to a subscripted element (such as x[10] or z[i][j][k]) causes *all* intermediate elements which do not already exist to be created and assigned the empty type, **Nothing**. This process is called **array-infilling**. Any newly created values are set to the value **nothing**. Array-infilling works for any number of indices and is a sharp sword.

In order to provide efficient support for manipulating time series, SUB applies special rules to certain types of array manipulations. The rules are are simple, but they require some understanding of what SUB thinks it is being told to do.

B.3.2 Simple Assignment

A simple SUB assignment statement

```
a = b;
```

stuffs a copy of whatever is in b into a. This new value completely obliterates whatever was in a before the assignment.

> **Exception**
>
> If a is actually a record member with some particular nature, such as x.trace when x holds a SEG-Y trace record, SUB knows it must preserve the member's character; we can't replace a trace with a string, for example. In this instance it will do its best to find some sort of assigment that makes sense and do it (or die trying).

Assigning into an array element works the same way, subject to *array-infilling*. This

```
ay[5] = 33;
```

causes ay[5] to be created if necessary and its contents set to 33. Array-infilling makes sure that

```
ay[0], ay[1], ..., ay[5]
```

exist. Any newly created values are set to the value **nothing**.

B.3.3 Subranges

If a is a SUB array, then

— a denotes the entire array,

— a[2] denotes a single element of the array, and

— a[3:28] denotes a sub-array of the entire array.

We support default values for the two indices that specify a subrange. If the first index is missing, it's value defaults to 0. If the second index is missing, it's value defaults to the value of the last index in the array.

Subranges are useful both as values (on the right side of an assignment statement) and as targets (on the left side of an assignment statement).

Value of a Subrange. When we *evaluate* a subrange, it's value is just the array formed by the elements in the subrange. Note that the sub-array's indices start at 0, as in this example

```
x = z[3:7];
//      now      x[0] == z[3]
//               x[1] == z[4]
//      etc.
```

Assignment to a Subrange. When we assign a value to a subrange, the rules change. If we assign a scalar into a subrange,

```
x[12:98] = 3.142;
```

the value of the scalar (here 3.142) is stored into each member of the subrange. If z is an array with ten values, each of these expressions

```
z[:]    = 0.0;
z[0:9]  = 0.0;
z[0:$]  = 0.0;  // note that $ works here
z[:9]   = 0.0;
```

sets all of the members of z to zero.

If we assign an array into a subrange, the rules are a little different. In this case, if the array is shorter than the subrange to which it is being assigned, the array is extended *by adding zeroes* until it is long enough. Note that we **do not** replicate the last value in the array. Here's an example:

```
//      create an array with the values 1, 2, 3, 4, 5.
```

```
z = sequence(1,5);

//      create array with 12 entries and set them all to 1.

a = sequence(1,12);
a[:] = 1;

//      a now has twelve members, each holding the value 1.
//
//      now assign into the 7-element
// subrange a[2:8]

a[2:8] = z;

//      here's what a has now:
//          a[0] = a[1] = 1         unchanged
//          a[2] = 1,...            from z
//          a[6] = 5                from z
//          a[7] = 0                from z's extension
//          a[8] = 0                from z's extension
//          a[9] = ... = a[11] = 1  unchanged
```

B.3.4 fVectors

SUB has two different types of arrays. Conventional arrays can hold any collection
of SUB values, including other arrays. When you create an array by assignment,

```
b[22] = ''help!'';
```

you create a conventional array.

An fVector is a special type of array that contains only floating-point values.
These arrays can only be created by assignment (the library contains functions
to create new ones for you) or by accessing one in a data record. These arrays
obey the normal rules for SUB arrays but they also offer support for high-level
numerical manipulation.

Manipulating fVectors. Currently fVectors understand how to add, subtract,
multiply, and divide with either another fVector or a scalar. Each of these operations is executed component-wise. Here are some examples:

```
mytr = Tr.trace;
//
```

```
//      mytr now holds an fVector
//
//      this will compute it's mean value and then subtract
// it from the trace.
//                                                      ʼ
meanv = sum(mytr) / length(mytr);
mytr = mytr - meanv;
//
//      note that we subtracted a scalar (meanv) from a
// trace (mytr) producing a new trace with mean zero.
//
//      Here's an estimate of the integral over [0,1] if
//              sqrt(exp{y(t)}) dt
//      give or take a few scale factors, using the vector
// capability of the transcendental functions.
//
estimated_integral = sum(sqrt(exp(mytr))) / size(mytr);
//
//      etc.
```

The library functions fVector, sequence, and flatten are useful for creating fVectors.

B.4 SUB Library

B.4.1 Transcendental and Elementary Functions

Most of these are defined in the obvious way. Trigonometric functions exect arguments in radians (see the appropriate section 3 manual page for details).

Functions which claim to be *vectorized* can be applied to multi-dimensional arrays of arbitrary (!) complexity subject to the condition that each end-member of the array is a number. (In other words, every element of the argument must be either a number or another array and this requirement holds recursively.) The return value is a new array of similar shape in which all of the original numeric values have been replaced with the results of applying the function to them.

sin(x)	vectorized
cos(x)	vectorized
atan(x)	vectorized
log(x)	vectorized
log10(x)	vectorized
exp(x)	vectorized
sqrt(x)	vectorized
int(x)	vectorized; discards any fractional part
nint(x)	vectorized; rounds to the nearest integer
abs(x)	vectorized

B.4.2 Arithmetic Functions

— max($a,b,c,...$)

Returns the algebraic maximum of all of the numeric values in all of the (arbitrarily complex) arguments. There must be at least one argument.

— min($a,b,c,...$)

Like max but it returns the algebraic minimum.

— sum($a,b,c,...$)

Like max but it returns the sum of all of the numeric values.

B.4.3 Array Construction and Management Functions

— flatten($a,b,c,...$)

Returns a one-dimensional array constructed by flattening all of its arguments. Each argument can be either a single value or an array of arbitrary complexity. The original arrays are not altered (since we have only call-by-value in SUB).

— sequence(*number*)

— sequence(*from, to*)

— sequence(*from, to, step*)

Each of these calls returns an initialized fVector. The first form returns a vector initialized to the values

$$1, ..., number$$

The second returns a vector initialized to

$$from, \ from \ + \ 1, ..., \ to$$

The third returns a vector initialized to

$$from, \ from \ + \ step, ..., \ last$$

where *last* is the final element in the sequence which is not greater than *to*

— fvector(*i*)

— fvector(*i, c*)

Returns an fVector with *i* elements. If an optional second argument, *c*, is provided, the array elements are initialized to that value, otherwise they are initialized to 0.

B.4.4 Time Series Analysis

Time series analysis is currently limited to forward and inverse digital fourier transform capability. (Our *dft*'s are based on routines written by Dave Hale, and available from the Center for Wave Phenomena.) Here's what we have:

— pfnext(*n*)

Our fourier transform code is limited to series lengths with certain properties (see the source code for more on this). (It is not limited to just powers of two, however.) This function accepts an arbitrary series length, *n*, and returns the first *acceptable* length which is not less than *n*. We provide another function, extend() below, which can be called to pad a series to the desired length.

— pfnext(*n, nmax*)

This function accepts an initial series length, *n*, and a maximum acceptable length, *nmax*, for an extended series and returns an acceptable series length which is at least as large as *n*, is no larger than *nmax*, and which we believe (based on timing tests conducted by the dft code's author) will provide the most eficient calculation.

— extend(*y, n* [,*val*])

Returns a new series derived from *y* which has *n* elements. If the new series is longer than *y*, the new elements are initialized to *val* if *val* is provided, and 0 otherwise.

— fdft(*y*)

Accepts an input time series, which may be real or complex (see below), and returns its fourier transform. The returned value consists of a two-element array; the first element of the array is an **fVector** which is the *real* part of the transform, and the second element of the array is the *imaginary* part of the transform.

The input series, *y*, can be either an **fVector**, in which case it is taken to have no imaginary component, or it can be a two-element array each of the components of which is an array of real values, in which case the arrays are interpreted just as in the above. (Both **fVectors** and conventional arrays are acceptable.)

− idft(*y*)

Provides the inverse of **fdft()**. It's behavior is identical to that described above except that the inverse transform is computed and the results are divided by the number of samples in the series, so that the combination of a forward and an inverse transform is the identity operation.

Check out the examples given in a later section.

B.4.5 Other Numeric Vector Stuff

There are a few other functions intended to support processing vectors of numeric values.

− sort(*y*[, *descending*])

Returns a sorted copy of *y*. More precisely, if *y* cannot be flattened into an **fVector**, **sort** returns an empty value; otherwise it returns an **fVector** produced by sorting a flattened version of *y*. By default *y* is sorted in increasing order. If the optional argument *descending* is present and non-zero, the sense of sorting is reversed.

− reverse(*y*)

Returns a reversed flattened copy of *y*. The argument does not have to be entirely numeric.

− bin(*y*, *bins*[, *lowest*, *highest*])

Returns a binning of a flattened copy of *y* (which must be numeric). Binning is into *bins* non-overlapping, contiguous bins. By default the first bin starts at *y*'s minimum value and the last bin ends at *y*'s maximum value. If the optional arguments *lowest* and *highest* are present, those arguments determine the binning limits. (This stuff is all John Scales' fault.)

B.4.6 Function Argument Handling

SUB provides two library functions that allow a user-defined SUB function to determine the number and value of its arguments.

– nargs()

Returns the number of arguments passed to the current function.

– nthargs()

Returns the value of the *n*th argument. ntharg(0) is the name of the current function; ntharg(1) is the value of the first argument. It is an error to ask for arguments that aren't there.

B.4.7 Command Line Arguments

The SUB library currently lacks slick support for parsing command-line arguments (the stuff on the command line after the name of the script file). We do, however, provide two global variables:

– argc()

Contains the number of command line arguments, counting the name of the script file as the first. This value is always greater than 0.

– argv()

Is an array of strings holding the values of the command line arguments. argv(0) is the name of the script file and argv(1) is the first subsequent argument (if any).

B.4.8 Type Predicates

These are functions that provide information about the type (as opposed to the value) of the current contents of a variable. We don't currently cover all possible types (but we could - it's just not a high priority).

— isDouble(*arg*)

returns true (1) if *arg* currently holds a value of type Double.

— isString(*arg*)

— isNothing(*arg*)

— isArray(*arg*)

— isfVector(*arg*)

B.4.9 Input/Output

Formatted I/O.

— fgets(*stream*)

Reads characters from *stream* until end-of-file or a newline is encountered. Returns a string with all of the characters read (including a terminating newline if one was encountered).

— print(*a,b,c,...*)

Writes formatted forms of the arguments *a,b,c,etc* to stderr.

— fprint(*stream, a,b,...*)

Writes formatted forms of *a,b,c,   etc* to *stream*.

Formatted I/O.

— getTokenLine(*stream*)

— getTokenLine(*stream, separators*)

The first form of this function reads the next line of input from *stream*, breaks it into tokens (see below), and returns an array with one element for each token. On end-of-file, this function returns the value Nothing.

A token is normally any set of characters bounded by whitespace (spaces, tabs, and newlines). In the second form, the token separator characters are taken to be any characters in the string *separators*.

Record I/O.

— fputrb(*obj1, obj2, ...*)

— fputrb(*stream, obj1, obj2, ...*)

Sends each of the records *obj1*, *obj2*, ..., to stdout. As a special case, if the first argument is a stream (such as the value returned from fopen()), that stream replaces stdout as the destination of the write.

— output(*obj*)

A synonym for fputrb which is allowed for historical reasons. This function will be removed one day.

— fgetrb()

— fgetrb(*stream*)

Reads the next record from *stream* (stdin by default) and returns it. Returns Nothing on end-of-file.

Opening and Closing Files.

— fopen(*filename, mode*)

Opens the file specified by the path *filename* for i/o in the direction specified by the string *mode* and returns a file pointer value suitable for passing to fprint (below). *Mode* should be one of "r", "w", "a", "r+", "w+", or "a+" where the quotes are required (the first two are by far the most common); see fopen(3) for more details.

— tmpfile()

Returns a *stream* opened in update mode to a temporary file which will be automatically deleted when the file is closed or the current process terminates.

— fclose(*stream*)

Closes a *stream* (which must be a value returned by an earlier call to fopen).

— fflush(*stream*)

Writes any buffered data to *stream*.

Opening and Closing Pipes.

— popen(*cmd, mode*)

Opens a pipeline to the process *cmd* in the direction (reading or writing) specified by *mode*, and returns a file pointer value suitable for passing to fprint, *etc. Cmd* is executed to create the target process and can be any legal sh(1) command string. *Mode* should be "w" for writing to the *cmd* and "r" for reading from the *cmd* (the quotes are required).

– pclose(*stream*)

Closes *stream* (which must be a value returned by an earlier call of popen()). This call closes the i/o stream and waits for the remote process to exit (see popen(3)).

Manipulating Streams. These functions support various types of manipulation of I/O streams. Most of them correspond closely to entries in the Unix manual section 3.

– fseek(*stream, offset, mode*)

Positions *stream* to a new location determined by *offset* and *mode*. *Mode == 0*, new location is at *offset* measured from the beginning of the stream. *Mode == 1*, new location is at *offset* measured from the current location. *Mode == 2*, new location is at *offset* measured from the end of the stream.

– ftell(*stream*)

Returns the current offset of the stream from its beginning.

– rewind(*stream*)

Equivalent to fseek(*stream*,0,1)

– fskip(*stream,   bytes*)

Skips *bytes* on *stream. Stream* must be opened for input.

Bytewise I/O.

– fgetbytes(*stream, n*)

Reads up to *n* bytes from *stream*, which must be opened for input. If any bytes are read, this functions returns an array with one numeric element for each byte. Otherwise it returns the value *nothing*.

– fputbytes(*stream, a*)

Writes one byte to *stream* for each element in the array *a. Stream* must be opened for output and *a* must be an array of numeric values.

B.5 Miscellaneous Functions

– time()

Returns the current wall clock time in (double) seconds since 0:00 GMT, January 1, 1970. Granularity is system dependent (see gettimeofday(2)).

– random()

– random(n)

returns a random number in the half-open interval [0,1); if the optional argument n is provided, returns an fVector with n random elements. (The current random number generator is based upon freely available code discussed in the source files.)

– exit(x)

Causes SUB to teminate and return the value x to the shell that invoked it. By convention exit(0) denotes success and any other value indicates failure. A script which exits by falling off the end returns 0.

– system(s)

Executes the shell command s, which must be a string, waits until the command has completed, and then returns the exit status of the shell which invoked the command. See system(3).

– strlen(s)

Returns the number of characters in the string s (actually a synonym for size()).

– size(x)

Returns the size of x (this might be useful in detecting bad header sizes, *etc*). If x is a string the returned value is the length of the string (the trailing nul is not counted). If x is an SEG-Y trace record the returned value is the size of the record in bytes. If x is an array (either conventional or fVector), the returned value is the number of elements. If x has the type Nothing, the returned value is 0. In all other caes, the returned value is 1.

– strtonum(s)

Interprets the contents of s (which had better be a string) as a number and returns its value. If s does not at least begin with a legitimitate numeric value, this function will abort with an error message.

– floattostr(*f, fmt*)

Converts the value of *f*, as a floating-point number, into a string using the printf(3) format in the string *fmt*. *Fmt* should contain a floating-point format string such as "%g", *etc.*

– inttostr(*i, fmt*)

Converts the value of *f*, as n integer, into a string using the printf(3) format in the string *fmt*. *Fmt* should contain an integer format string such as "%d", *etc.*

B.5.1 Predefined Values

We provide a set of constant values accessible through global variables.

– pi

3.14159265358979323846

– enatural

2.71828182845904523636

– gamma

0.57721566490153286060 (Euler's constant - for the snobs in the audience)

– rad2deg

57.29577951308232087860 (degrees per radian)

– golden

1.61803398874989484820 (the golden mean - for the esthetes)

– nothing

The empty value, a pile of ashes

– hardware

A string name for the hardware upon which we are running (such as "sun", "cray").

B.5.2 Debugging Support

— traceback()

Causes the current function stack, including the values of arguments on the stack, to be printed to **stderr**. Execution resumes after the values are printed.

— internals(*selector*)

Returns the values of various internal quantities of interest during debugging of the SUB interpreter. These are surely of no interest to anyone else, but here they are: internals(0), the current values of sbrk(2) and internals(1), the current alue of nextTemp from opcode.c.

B.6 Accessing Data Records

One of SUB's design goals is to be readily adaptable to new data formats (that is, more than just SEG-Y). Because of this we sometimes refer to SEG-Y records as RBlocks. In the SEG-Y version, which you are using now, rBlocks are exactly the same thing as SEG-Y trace records.

B.6.1 Member Access

The RBlock type supports a structure-like convention for access to individual data elements (including the trace data samples). If, for example, rb holds a trace record, then rb.ns provides (read or write) access to the number-of-samples field.

In addition to the simple header fields, the trace samples can be accessed through a pseudo-array. If rb holds a trace record, rb.trace[*i*] will access the *i*th sample.

All of the member names and meanings for SEG-Y trace records are summarized in the table at the end of this appendix.

It is possible to change the size of the trace by assigning a new fVector to it. Thus

```
Tr.trace = fvector(10, 0);
```

would replace the original trace with a trace with ten samples, all with the value 0. Whenever a length-changing assignment is made to a trace, SUB attempts to set the value of the member **ns** to correspond to the new length. If you want **ns** to have a value other than the current trace length, you must assign to ns *after* changing the trace.

```
//   suppose z is a SEGY record and z.trace has 20 elements.
//   initially z.ns will have the value 20.
z.ns = 7;                     // z.ns now has the value 7
z.trace = fvector(12, 0);  // z.ns now has the value 12
z.ns = 7;                     // z.ns is seven again
```

B.7 Programming

This is a collection of specialized discussions, tips, and examples.

B.7.1 Pipes

Popen() and pclose() provide very substantial flexibility to scripts. A child process which is popen'd in mode "w" shares stdout with the parent process (sub in this case). Thus a child process invoked in this manner can be used as an output filter for the parent process. Values written by the child process to its stdout will emerge *as though* from the parent process' stdout. A similar mechanism works for read pipelines.

Remember that many programs (such as xgraph(1) which is used in one of the examples) do not do anything interesting until they have read everything available from stdin. These programs will not see end-of-file on stdin until the SUB script has called pclose().

B.7.2 A Plotting Example

This script reads the first five traces in a SonicTool data set, computes the amplitude spectrum of each, and displays the results using a pipe to the public-domain plotting tool xgraph. (**NB:** this **xgraph** is not to be confused with the **xgraph** that comes with **SU**. They are not the same. Another good free-ware plotting tool is **xvgr. Xvgr** reads from standard in when given the command line argument **-source stdin.**)

```
//  this script assumes that the sample interval is
//  not stored in the SEGY record; it's set to 10 uS
//  in Begin().  If the sample interval is stored in
//  the header you can extract from there.  We don't
//  plot the entire amplitude spectrum since it's not
//  usually interesting over its entire range.

func Begin() {
    dt = 10.0e-6; // assume dt = 10 uS
    fmin = 500;   // minimum frequency to display
    fmax = 10000; // maximum frequency to display
    Records = 0;
xg = popen("xgraph", "w");
}

func plotAmplitude(series) {
    auto et, ftet, pet, nx, i, df, f;
    df = 1.0 /(dt * size(series));
    et = extend(series, pfnext(size(series)));
    ftet = fdft(et);
    pet = sqrt(ftet[0]*ftet[0] + ftet[1]*ftet[1]);
    nx = size(pet)/4;
    fprint(xg, "\\n");
for(i = 0; i < nx; i++) {
        f = i * df;
        if((f >= fmin) && (f <= fmax))
            fprint(xg, f, " ", pet[i], "\\n");
}
}

func OnTrace() {
    ++Records;
    plotAmplitude(Tr.trace);
    if(Records < 5) return;
    fclose(xg); // don't wait for end of data set.
    exit(0);    // wrap up and exit now.
}
```

B.7.3 Vectors Versus Loops

If efficiency is an issue (and usually not important for many SUB applications), vectorized arithmetic is a lot faster than wading through a loop. On the other hand, loops are more flexible.

Here are two scripts that perform the same simple calculation. The first one uses a loop:

```
//
//   20 passes by 1000 elements takes 19.2 sec of cpu time
// about 1 second/pass
//
func computeL2(lv) {
    auto i, s;
    s = 0;
    for(i = 0; i < size(lv); i++)
     s = s + lv[i];
    return s;
}

func Begin() {
    auto i;
    lv = random(1000);
    for(i = 0; i < 20; i++)
     computeL2(lv);
    exit(0);
}
```

The second one uses vector operations:

```
//
//   200 passes by 1000 elements takes 0.74 sec of cpu
//   time about 3.7 milliseconds/pass.
//
func computeL2(lv) {
    return sum(lv*lv);
}

func Begin() {
    auto i;
    lv = random(1000);
```

```
        for(i = 0; i < 200; i++)
          computeL2(lv);
        exit(0);
    }
```

As the timing information in the examples shows, the vectorized version was about *300* times faster.

Table B.1. SEG-Y Headers – Part I

	SEGY Trace Record Members		
type	name	length	meaning
integer	tracl	1	trace sequence number within line
integer	tracr	1	trace sequence number within reel
integer	fldr	1	field record number
integer	tracf	1	trace number within field record
integer	ep	1	energy source point number
integer	cdp	1	CDP ensemble number
integer	cdpt	1	trace number in CDP ensemble
short	trid	1	trace type identification code
short	nvs	1	number of vert. summed traces
short	nhs	1	number of horiz. summed traces
short	duse	1	data use: production (1) or test (2)
integer	offset	1	distance from source to rec. group
integer	gelev	1	rec. group elevation from sea level
integer	selev	1	source elevation from sea level
integer	sdepth	1	source depth (positive)
integer	gdel	1	datum elevation at receiver group
integer	sdel	1	datum elevation at source
integer	swdep	1	water depth at source
integer	gwdep	1	water depth at receiver group
short	scalel	1	scale factor for previous 7 entries
short	scalco	1	scale factor for next 4 entries
integer	sx	1	X source coordinate
integer	sy	1	Y source coordinate
integer	gx	1	X group coordinate
integer	gy	1	Y source coordinate
short	counit	1	coord. units code for previous 4 entries
short	wevel	1	weathering velocity
short	swevel	1	subweathering velocity
short	sut	1	uphole time at source
short	gut	1	uphole time at receiver group
short	sstat	1	source static correction
short	gstat	1	group static correction
short	tstat	1	total static applied
short	laga	1	lag time A (ms)
short	lagb	1	lag time B (ms)
short	delrt	1	delay recording time (ms)
short	muts	1	mute time–start
short	mute	1	mute time–end

Table B.2. SEG-Y Headers – Part II

\multicolumn{4}{c}{SEGY Trace Record Members}			
type	name	length	meaning
unsgn short	ns	1	number of samples in this trace
unsgn short	dt	1	sample interval, in micro-seconds
short	gain	1	gain type of field instruments code
short	igc	1	instrument gain constant
short	igi	1	instrument early or initial gain
short	corr	1	correlated 1 (no) or 2 (yes)
short	sfs	1	sweep frequency at start
short	sfe	1	sweep frequency at end
short	slen	1	sweep length in ms
short	styp	1	sweep type code
short	stas	1	sweep trace length at start in ms
short	stae	1	sweep trace length at end in ms
short	tatyp	1	taper: 1=linear, 2=$\cos^2$, 3=other
short	afilf	1	alias filter frequency if used
short	afils	1	alias filter slope
short	nofilf	1	notch filter frequency if used
short	nofils	1	notch filter slope
short	lcf	1	low cut frequency if used
short	hcf	1	high cut frequncy if used
short	lcs	1	low cut slope
short	hcs	1	high cut slope
short	year	1	year data recorded
short	day	1	day of year
short	hour	1	hour of day (24 hour clock)
short	minute	1	minute of hour
short	sec	1	second of minute
short	timbas	1	time base: local(1), GMT(2), other(3)
short	trwf	1	trace weighting factor
short	grnors	1	group number of roll sw posn one
short	grnofr	1	group number of trace one (orig)
short	grnlof	1	group number of last trace (orig)
short	gaps	1	gap (number of groups dropped)
short	otrav	1	overtravel taper code
float	d1	1	sample spacing, non-seismic data
float	f1	1	first sample loc., non-seismic data
float	d2	1	sample spacing between traces
float	f2	1	first trace location
float	ungpow	1	negative of power used for compress
float	unscale	1	reciprocal of range scaling factor
short	mark	1	mark selected traces
short array	extra	17	unassigned
float array	trace	?	trace samples

References

1. K. Aki and P. G. Richards. *Quantitative seismology: theory and practice*. Freeman, 1980.
2. R. M. Alford, K. R. Kelly, and D. M. Boore. Accuracy of finite-difference modeling of the acoustic wave equation. *Geophysics*, 39:834–842, 1974.
3. R. G. Bartle. *The elements of real analysis*. Wiley, 1976.
4. E. Baysal, D. D. Kosloff, and J. W. Sherwood. Reverse time migration. *Geophysics*, 48:1514–1524, 1983.
5. J. Berryman. Lecture notes on nonlinear inversion and tomography, 1991.
6. N. Bleistein and R. A. Handelsman. *Asymptotic expansions of integrals*. Dover, 1986.
7. M. Båth. *Mathematical aspects of of seismology*. Elsevier, 1968.
8. M. Born and E. Wolf. *Principles of optics*. Pergamon, 1980.
9. C. Brezinski. *History of continued fraction and Padé approximants*. Springer-Verlag, 1980.
10. H. B. Callen. *Thermodynamics*. Wiley, 1960.
11. V. Cerveny. Ray tracing algorithms in three-dimensional laterally varying layered structures. In Guust Nolet, editor, *Seismic tomography*. Reidel, 1987.
12. C. Chapman. A new method for computing synthetic seismograms. *Geophysical Journal R. Astr. Soc.*, 54:481–518, 1978.
13. C. Chapman. Generalized Radon transforms and slant stacks. *Geophysical Journal R. Astr. Soc.*, 66:445–453, 1981.
14. J. Claerbout. *Imaging the Earth's interior*. Blackwell, 1986.
15. R. Clayton and B. Engquist. Absorbing boundary conditions for the acoustic and elastic wave equations. *Bulletin of the Seismological Society of America*, 67:1529–1540, 1977.
16. A. M. Correig. On the measurement of body wave dispersion. *J. Geoph. Res.*, 96:16525–16528, 1991.
17. M. B. Dobrin and C. H. Savit. *Introduction to geophysical propsecting*. McGraw Hill, 1988.
18. P. C. Docherty. Kirchhoff migration and inversion formulas. *Geophysics*, 56:1164–1169, 1991.
19. M. Ewing, W. S. Jardetzky, and F. Press. *Elastic waves in layered media*. McGraw Hill, 1957.
20. K. Fuchs and G. Müller. Computation of synthetic seismograms with the reflectivity method and comparison with observations. *Geophysical J. Royal Astr. Soc*, 23:417–433, 1971.
21. J. Gazdag. Wave equation migration with the phase-shift method. *Geophysics*, 43:1342–1351, 1978.
22. J. Gazdag and P. Sguazzero. Migration of seismic data. *Proceedings of the IEEE*, 72:1302–1315, 1984.

23. C. W. Gear. *Numerical initial value problems in ordinary differential equations.* Prentice-Hall, 1971.
24. I. M. Gel'fand and G. E. Shilov. *Generalized functions: Volume I.* Academic Press, 1964.
25. H. Goldstein. *Classical Mechanics.* Addison-Wesley, 1950.
26. S. H. Gray. Efficient traveltime calculations for kirchhoff migration. *Geophysics*, 51:1685–1688, 1986.
27. N. B. Haaser and J. A. Sullivan. *Real analysis.* Van Nostrand Reinhold, 1971.
28. C. Hemon. Equations d'onde et modeles. *Geophysical Prospecting*, 26:790–821, 1978.
29. M. Hestenes. *Conjugate direction methods in optimization.* Springer-Verlag, Berlin, 1980.
30. J. D. Jackson. *Classical Electrodynamics.* Wiley, 1975.
31. F. John. *Partial differential equations.* Springer-Verlag, 1980.
32. D. S. Jones. *Acoustic and electromagnetic waves.* Oxford University Press, 1986.
33. H. Keller. *Numerical methods for two-point boundary value problems.* Blaisdell, 1968.
34. K. R. Kelly, R. W. Ward, S. Treitel, and R. M Alford. Synthetic seismograms: a finite-difference approach. *Geophysics*, 41:2–27, 1976.
35. R. Langan, I. Lerche, and R. Cutler. Tracing rays through heterogeneous media: an accurate and efficient procedure. *Geophysics*, 50:1456–1465, 1985.
36. A. R. Mitchell. *Computational methods in partial differential equations.* Wiley, 1969.
37. P. M. Morse and H. Feshbach. *Methods of theoretical physics.* McGraw Hill, 1953.
38. R. D. Richtmyer and K. W. Morton. *Difference methods for initial value problems.* Interscience, 1967.
39. E. Robinson, T. S. Durrani, and L. G. Peardon. *Geophysical signal processing.* Prentice Hall, 1986.
40. J. A. Scales and M. L. Smith. *Introductory geophysical inverse theory.* Samizdat Press, 1994.
41. W. A. Schneider. Integral formulation of migration in two and three dimensions. *Geophysics*, 43:49–76, 1978.
42. G. D. Smith. *Numerical solution of partial differential equations: finite difference methods.* Oxford, 1978.
43. A. Sommerfeld. *Optics.* Academic Press, 1964.
44. J. Stoer and R. Bulirsch. *Introduction to numerical analysis.* Springer-Verlag, 1980.
45. D. Struik. *Lectures on classical differential geometry.* Addison-Wesley, 1950.
46. A. Tarantola. *Inverse Problem Theory.* Elsevier, New York, 1987.
47. P. Temme. A comparison of common-midpoint, single-shot, and plane-wave depth migration. *Geophysics*, 49:1896–1907, 1984.
48. S. Treitel, P. Gutowski, and D. Wagner. Plane-wave decomposition of seismograms. *Geophysics*, 47:1372–1401, 1982.
49. J. Vidale and H. Houston. Rapid calculation of seismic amplitudes. *Geophysics*, 55:1504–1507, 1990.
50. G. N. Watson. *A treatise on the theory of Bessel functions.* Cambridge University Press, 1962.

Index

Lecture Notes in Earth Sciences

Vol. 37: A. Armanini, G. Di Silvio (Eds.), Fluvial Hydraulics of Mountain Regions. X, 468 pages. 1991.

Vol. 38: W. Smykatz-Kloss, S. St. J. Warne, Thermal Analysis in the Geosciences. XII, 379 pages. 1991.

Vol. 39: S.-E. Hjelt, Pragmatic Inversion of Geophysical Data. IX, 262 pages. 1992.

Vol. 40: S. W. Petters, Regional Geology of Africa. XXIII, 722 pages. 1991.

Vol. 41: R. Pflug, J. W. Harbaugh (Eds.), Computer Graphics in Geology. XVII, 298 pages. 1992.

Vol. 42: A. Cendrero, G. Lüttig, F. Chr. Wolff (Eds.), Planning the Use of the Earth's Surface. IX, 556 pages. 1992.

Vol. 43: N. Clauer, S. Chaudhuri (Eds.), Isotopic Signatures and Sedimentary Records. VIII, 529 pages. 1992.

Vol. 44: D. A. Edwards, Turbidity Currents: Dynamics, Deposits and Reversals. XIII, 175 pages. 1993.

Vol. 45: A. G. Herrmann, B. Knipping, Waste Disposal and Evaporites. XII, 193 pages. 1993.

Vol. 46: G. Galli, Temporal and Spatial Patterns in Carbonate Platforms. IX, 325 pages. 1993.

Vol. 47: R. L. Littke, Deposition, Diagenesis and Weathering of Organic Matter-Rich Sediments. IX, 216 pages. 1993.

Vol. 48: B. R. Roberts, Water Management in Desert Environments. XVII, 337 pages. 1993.

Vol. 49: J. F. W. Negendank, B. Zolitschka (Eds.), Paleolimnology of European Maar Lakes. IX, 513 pages. 1993.

Vol. 50: R. Rummel, F. Sansò (Eds.), Satellite Altimetry in Geodesy and Oceanography. XII, 479 pages. 1993.

Vol. 51: W. Ricken, Sedimentation as a Three-Component System. XII, 211 pages. 1993.

Vol. 52: P. Ergenzinger, K.-H. Schmidt (Eds.), Dynamics and Geomorphology of Mountain Rivers. VIII, 326 pages. 1994.

Vol. 53: F. Scherbaum, Basic Concepts in Digital Signal Processing for Seismologists. X, 158 pages. 1994.

Vol. 54: J. J. P. Zijlstra, The Sedimentology of Chalk. IX, 194 pages. 1995.

Vol. 55: J. A. Scales, Theory of Seismic Imaging. XV, 291 pages. 1995.

Vol. 56: D. Müller, D. I. Groves, Potassic Igneous Rocks and Associated Gold-Copper Mineralization. XIII, 210 pages. 1995.